The Crustacean Codex

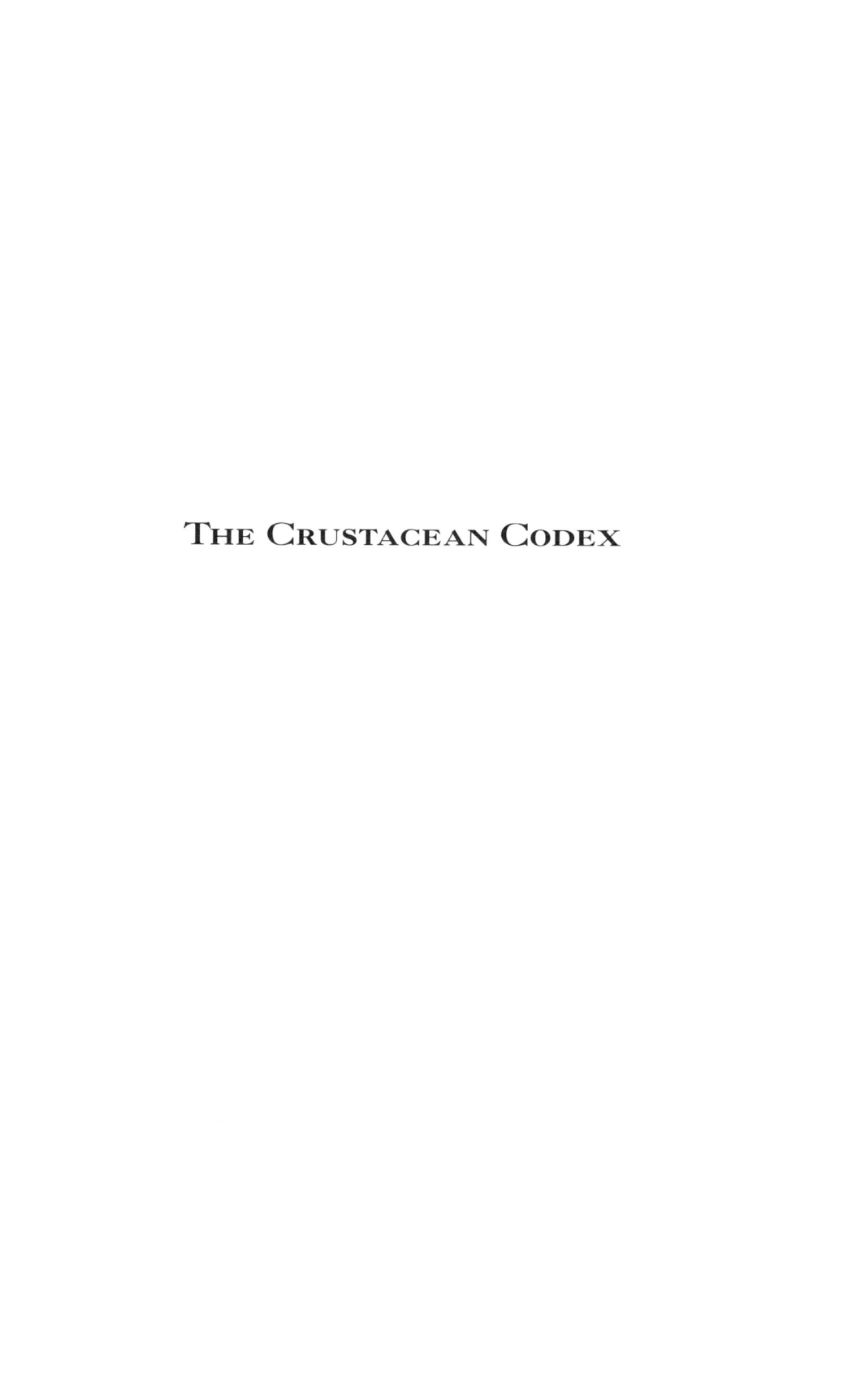

The
Crustacean Codex

Thomas Suárez

Terra Nova Press
1997

Copyright © 1997. This book is copyright by Thomas Suárez. All rights reserved. Except as permitted under the Copyright Act of 1976 and amendments thereto, no part of this book may be reproduced in any form or by any electronic or mechanical means, including the use of information storage and retrieval systems, or any digital technology, without permission in writing from the copyright holder. Requests for permissions should be addressed in writing to: G.B. Manasek, Inc. Box 1204, Norwich VT 05055-1204 USA.

ISBN 0-9649000-5-X

First edition. First printing. 1997.

TERRA NOVA PRESS
Norwich, VT 05055-1204
USA

for Sainatee

Author's Note :

While writing the book I frequently consulted with my friend Richard Casten for his wise insights (it is my good fortune that we live in the same universe). Another smart friend, Kristina Nilsson, meticulously scrutinized the original draft, and her comments contributed greatly to the final text. Gina Suárez and John Suárez were both of enormous help with proofreading and suggestions. Frank Manasek of Terra Nova Press provided expertise – and patience.

And my wife Ahngsana, as always, kept our twig of the universe filled with beautiful song.

This is also a fitting place for me to thank my parents for having made sacrifices to allow me an education in music when I was young — aside from, of course, thanking them for everything else imaginable.

— T. S.

THE CRVSTACEAN CODEX

being a

TRVE CHRONICLE

and

COMPLETE HISTORIE

of the

VNIVERSE

through the Present Yeare of our Lore.

Pliny Planktonius, auctore

URTEXT

Chiengmaius, apud Tomaso Suarezius
Ianuarius 2, anno mcmxciii

Contents

	Foreword	11
I.	Prologue	13
II.	A Message In The Reef	15
III.	Tea - Time	26
IV.	Journey To The Middle	42
V.	Passage To Terra Unherded	64
VI.	Zinnia	78
VII.	Zero Degrees On All 'Tudes	90
VIII.	Fugafest	110
IX.	The Sea Of Kopólo	127
X.	Epilogue	140

Foreword

The anonymous manuscript which has come to be known as the *Crustacean Codex* is believed to be based largely on the lost *Historie* of Pliny Plankton the Elder. Some parts, however, have clearly been derived from other sources, possibly reconstructed from oral tradition by the unknown scribe who produced the sole surviving manuscript.

The editor has taken the liberty of annotating the *Codex* with an occasional footnote where its meaning or context may not be sufficiently self-explanatory.

— T.S.

I
Prologue

> There
> once just
> happened to be
> a universe. No one
> knew quite why or how
> it had come to be, perhaps
> because it was the only universe
> that anyone in it had ever known. This
> universe had ocean breezes and misty mountains
> and water buffalo and gray silky clouds and all sorts of
> other caprices, both known and unknown to its various
> inhabitants, both named and unnamed in their various
> tongues. All its parts interacted with each other, how-
> ever elusively, and thus in a sense combined to form
> a single magnificent Being. For those who were a
> part of it, this wasn't just any universe
> — it was *the* Universe.

This Universe was truly a universe *a camera*, not *a regnum*. It knew neither homage to any master nor adulation of any baton. As a result, its marvelous ensemble was not always evident. Galaxies, in particular, gave the appearance of ignoring each other, only the rarest of observers being privy to their very subtle chamber music. And the various components of galaxies were almost as badly misunderstood, for the brilliance of their intricate counterpoint eluded most people.

It was only when narrowing one's view to solar systems that some sort of order became readily apparent. Yet, even here, many were utterly blind to the true poetry of their deft dance, perceiving solar systems to be clans of planets mindlessly fulfilling vows of submission to stars, moons relentlessly orbiting planets out of witless habit, meteors fatalistically awaiting their destiny at the atmospheric stake, and a few supervisory comets spying on the whole affair.

But such was hardly the case.

[Note regarding the first two sentences of the Prologue —]

It is commonly agreed among scientists that the universe is expanding. Many physicists believe that the universe as we know it, with advanced life and cognizant creatures such as ourselves, could exist only if it is expanding at a very *specific rate*, rather than any random rate. How the universe could have attained that one precise rate of expansion has been the point of considerable contention.

One theory, the "anthropic principle," suggests that there was neither any coincidence nor mechanism through which this unique rate of expansion was achieved, but rather that there are an undetermined number of other universes co-existing with ours, or many regions of a single large or infinite universe, each limited in space and time, each expanding at its own rate. According to this theory, the only universe (or region) with the proper conditions to produce us, creatures able to investigate such a question in the first place, would be that with the correct rate of expansion; such life could not have come to be in other universes or regions.

Thus (or in any case), as the beginning of the *Codex* observes, our Universe just happened to be, and it is the only universe that any of us have ever known. — ED.

II
A Message In The Reef

In an infinitesimally small corner of this Universe there was a particular galaxy in which there was a particular solar system in which there was a particular planet called Earth. In the most remote corner of this planet there was an ocean, called the Aquanesian Sea. This ocean had many reefs, most of which harbored shrimp, among many, many other creatures. In one particular reef (it had no name) there lived two particular shrimp. Their names were Ishmael and Atollana.[1] As you can well imagine, these two shrimp were very, very obscure creatures indeed.

It is not precisely known when they lived, though estimates placing them in the Middle Crustacean Period, the era during which terra firmates (the two-legged ground-inhabiting human people) were rapidly increasing their presence in all parts of the Earth orb, are believed to be fairly accurate.[2]

This was a troubled era. The Aquanesian Sea was in a state of decay. Its reefs were becoming cloudy and even, some claimed, poisoned. Both Ishmael and Atollana, in fact, were orphaned at an early age by this havoc, their parents having perished when the Great Oil Plague devastated their reef after a terra firmate ship foundered nearby.[3]

---

[1] *Ishmael* was the Biblical outcast from, and adversary to, prevailing Society. *Atollana* apparently derives from *atoll*, i.e., 'atollana' would be a female inhbitant of an atoll or its reef. – ED.

[2] *Terra firma* = a landmass, hence *Terra firmates* = inhabitants of the dry land. The Middle Crustacean Period probably corresponds to the medieval period of *terra firmate* peoples, and should not be confused with the Cretaceous Period of 100 million years ago. – ED.

The Crustacean Codex

With poison in their waters came venom in their hearts. Fishfolk had come to distrust one another, and leaders of various fish factions fabricated hate with which to create fear to subdue their subjects. All agreed that their children's future was murky. But no fish appeared able to pin-point the source of their ocean's malady and remedy it. The only defense they knew against their ailing sea was to invoke the shield of their own obscurity, a philosophical tonic.

The practice of obscurity is generally associated with terra firmates (they are thought to have coined the term). Obscurity, however, was actually first practiced by shrimp. In ancient times, long before terra firmates walked the Earth, magnificent crustacean civilizations had been masters at the art of obscurity. In their gentle philosophy, crustacean classicists of ancient shrimpdom held that obscurity was a relative phenomenon and that it helped give a person perspective. But attitudes had changed. The prevailing contemporary wisdom in the Aquanesian Sea maintained that obscurity was quantifiable, and many fish used it merely to escape their troubles.

In the reef there lived a very old sea anemone who was Ishmael and Atollana's adoptive parent. She believed that all creatures, even the terra firmates, were quite equally obscure (or unobscure, depending on one's perspective); in her view, obscurity itself had been invented from the myth that beings were somehow separated from the whole of the Earth and the Universe. The very Earth, the old anemone was fond of musing, was itself quite obscure.

Neither the shrimp nor their fellow Aquanesian citizens, however, understood this. They feared the world outside their reef. It was the world beyond that had created their own perceived obscurity. The reef was the shrimp's *orbis aquarum*;

---

[3] Apparently a general term used to refer to many such catastrophes rather than a single event; not to be misunderstood as isolating blame to any one 'grand old party'. — ED.

A Message In The Reef

to wander off its coral ledge would be to swim off the edge of the Earth.

Convinced of their own insignificance, Ishmael and Atollana daydreamed of the enormousness of the ocean. Contemporary crustacean civilization, it should be noted, was provincial; but hidden deep in its collective subconscious were distant memories of earlier, ocean-faring days. Some shrimp still asserted that marvelous and exotic fishfolk lived in the distant seas, fishfolk bearing no similarity to any that reside within the familiar reef. Pliny Plankton the Elder, the great historian and scientist from days long gone, had spoken much about these dwellers of the far waters.[4] But such fragments were about all that shrimpdom knew of the rest of the world. Few shrimp had ever savored more than a murky glimpse at the dry earth beyond the sea, or at the terra firmates that walk its surface; still fewer had ever seen the winged creatures which were said to swim the ocean of air above. For the most part, all this was hearsay, hearsay based mainly on the reports of Pliny Plankton; for although shrimplore records the expeditions, long ago, of merchant crustaceans who set out in quest of other seas, none had returned to tell their story. The deteriorating state of life in Aquanesia, however, made some of its citizens wonder if answers to their plight might lie in the outside world. Perhaps there were still pristine waters not defiled with the insidious muck which was mysteriously suffocating so many seas. Perhaps there were oases free of the prejudices which infected Aquanesia and the reef. Such speculation sparked the fancies of many, but, as far as was known, the actions of none.

---

[4] Pliny Plankton the Elder is a corruption of *Pliny the Elder* (or *Caius Plinius Secundus*), Roman naturalist and historian, ca. 23 - 79 A.D. One work of Pliny survives, the *Historia Naturalist*, which describes the nature of the physical universe, as well as geography, anthropology, the arts, medicinal uses of plants, and other topics. Pliny perished investigating the great eruption of Vesuvius in 79 A.D. — ED.

The Necklace

The anemone who raised Ishmael and Atollana, a wise and majestic soul, was considered by the reef residents to be a guru of sorts regarding matters of natural science. As children, the two shrimp would sit under her guardian arms, and she would thrill them with tales of strange places in the far corners of the seas. These stories had been related to her by Pliny Plankton himself, whom she had known in her youth.

One gentle evening, Ishmael and Atollana were to rendezvous by La Vecchia (for that is what they called the ancient anemone) at twilight. Ishmael loved the golden red light of the tropical dusk, and tonight he felt particularly intoxicated by the sun's waning rays. As the sun-oriented reef fish were retiring for the evening, their lunaresque colleagues were preparing for the night's labors. Ishmael was fascinated by how various fish had figured life out so differently. He had a strange desire to compliment whomever was responsible for arranging daily shifts in such a clever way.

The sun had completely fallen when Ishmael reached the appointed place, though he could see Atollana with the light of the night's full moon. She was facing La Vecchia and was nearly still, her body gently tossing to the ocean's rhythmic whim. The amber lunar radiance was dancing on her as it glowed on the luminous anemone. Ishmael would have believed he had chanced upon a garden of oceanic delights, had Atollana not looked so troubled. As he approached, she threw her head around and spoke intensely. Something was very wrong.

"Oh, dear Ishmael, come quickly! La Vecchia is very ill, and wishes to speak to us both. Please hurry!"

Ishmael leaped through the water and joined Atollana. Now both shrimp faced the ailing aged anemone who, with some difficulty, began to speak. The surrounding reef hushed to a solemn silence.

"Listen to me, dear children," the anemone began, "my days have been long. But I know that this very evening will be my time to pass on. And before I die, I must make myself content. I must tell you that which I have never told anyone.

"Of all our wonderful chats, your favorites were my stories about Pliny Plankton the Elder, and what of this world he saw and told. As you know, I was Pliny Plankton's scribe. That was many, many years ago. Upon returning from an expedition Pliny would immediately come to me and recount all that he had seen and learned. Now, Pliny was a skittish fellow who constantly juggled several thoughts at once but I, because of my many limbs, could notate them all at the same time, in an organized fashion. After we arranged every word precisely as we wanted, I would make many copies of the finished narrative simultaneously, using each of my arms to make one copy. Pliny was delighted with that because, you must remember, terra firmate scribes are confined to writing only one copy at a time."

La Vecchia noticed a twitch in Ishmael's eyes at her last statement. "Not two?" he asked.

"No, dear Ishmael, even though terra firmates have two arms and two hands, they can execute only a single copy at a time. But do not think poorly of them on that account — their physiology is, like yours and like mine, a balance of considerations and adaptations. But back to the report — Pliny would soon have enough copies to distribute among the historians, scholars, intellectuals, and other interested souls of the reef. Once he had answered any questions that the report may have raised, he would be off again, away on some other adventure in the far seas. Here in the reef, I would keep the archives in good order and accessible to any who wished to study them. This was the routine we followed."

The anemone sighed, her failing arms trembling a bit. "That was such a long time ago," she reflected, "such a long, long time ago. It seems that I've outlived all my contemporaries on this reef.

"Well, there came to pass an extraordinary day. Something happened to Pliny. He had been away for several tides. Normally, when he returned from an expedition, we would immediately get busy writing all that there was to record. But this time, despite his lengthy absence, we did not write anything. Nothing at all. Instead, he revealed an object which he said he had found in his travels. The object was a necklace, a simple chain of gold which strung a clay-colored shell. There was a message, a poem of sorts, carefully etched on the shell. And this poem obsessed Pliny. It haunted him. He said it had been written by a terra firmate girl who had roamed the rivers and streams aimlessly, but whom he had never met.

"Nor, it seems, had any of the local fish known her, for it had been generations since the girl left it there. But the kindly local fish still cared for the necklace, checking on it now and then to be sure that it was safe and that it did not get buried in the sand. They told Pliny that the girl had left it to confide her thoughts with any soul who might truly care to know of her. But of this girl they knew nothing more. The fish considered themselves the necklace's caretakers, not its owners, and it was an ancestral judgement that any person who was truly inspired by it should become its new steward. So they gave it to dear old Pliny.

"When Pliny came back, he sat by himself on the edge of the reef and stared interminably at it. Before the next moon shone, he came to me once more, this time insisting that I keep the necklace, that I take over as its caretaker. He knew the poem intimately, and was going off in search of the resolution of its riddle. He felt that it was neither necessary nor fair for him to keep possession of the necklace itself."

La Vecchia paused to twist one of her limbs around a recess in a large rock whose base was anchored well below the ocean floor. Reaching into a dim cavity, she retrieved a rather flat, well-worn shell strung by a chain with a dull reddish-yellow gleam. She held it in front of the shrimp for a moment, but then concealed it in her arms to be sure they would not be distracted by it. Her words were, for the moment, more important.

"It was then that Pliny left for the last time. I don't know to where he headed. He mentioned a long-lost 'chanted paradise' — I assume he meant *en*chanted — known as the Kingdom of

A Message In The Reef

Prester Prawn. He didn't know quite where this Kingdom was, but rumor spoke of a mountain or gorge called Aqua-Abyss in a land known as Zinnia.[5] Pliny somehow felt that the answers to our reef's woes might be found in this Zinnia, indeed, that the necklace's cryptic message could lead him there and help remedy the mystery of our ocean's decay. He believed that we needed to see our world from a distance, from an entirely impartial perspective, and that this far realm would offer such a vantage. Pliny also believed, curiously, that he had discovered the location of the fabled Mar Kopólo.[6]

"But that was the last I ever saw of him. Now that I can no longer be the keeper of his necklace, I ask you to take it from me. Forgive me that I cannot explain to you what it means. And, my dear Ishmael and my dear Atollana, forgive me that . . . that I now choose to relinquish myself to eternity."

With that, La Vecchia gently folded her arms and died. In her last moments, she had carefully set Pliny's necklace on the ground in front of the two shrimp. As the movement of the water continually spread sand about, Ishmael reached down to retrieve it.

"Put this around your neck," he told Atollana, "for that is where it now belongs." Atollana nodded and put the necklace on.

The two shrimp wandered aimlessly along the edge of the nocturnal reef, eventually reaching a lone coral boulder at the frontier of open ocean. Erosion by fine sand had smoothed the coral surface to a slippery polish, but they found a shallow indentation in which they could comfortably sit while keeping a clear vista of the seascape. Atollana rested her head on

---

[5] *The Kingdom of Prester Prawn* is believed to be an allusion to *The Kingdom of Prester John*, a mythical Christian kingdom, long believed to lie in Africa in Abyssinia (hence *Aqua-Abyss* in *Zinnia*). Originally, however, the kingdom was believed to be somewhere in Asia, and its history is closely entwined with the Crusades. — ED.

[6] Presumably after the famous Venetian merchant (ca. 1254 - ca. 1324). — ED.

Ishmael's lap and softly cried at the loss of their guardian and friend. After a while Ishmael, though no less grieving over La Vecchia's death, tried to pull her from her gloom.

"Atol . . . ," he whispered to her, "do you not think that La Vecchia was willing to us some mission, some destiny, with her revelations of Pliny Plankton and the shell? I believe she thought this necklace to be very important. It is now in our custody and trust. Come. Let's read its poem."

Atollana, aware of how silly La Vecchia would have thought them to despair so over her death, reached for the necklace. She angled the shell to catch the moonlight, and read its poem aloud while Ishmael looked on.

It said :
Find the point from which we spin
and sing a song to please my twin.
  But heed the fate
  that will await
  where perfect peace remains within :
Whisk ye to where the dew looms,
'tis dew of bloom that life consumes.
I wandered where,
where wandered I,
I'll wander to where none ask why,
and set upon my tears to dry.
      — *Honua*[7]

The two shrimp sat silently for a while, and then read the poem once more. They tried to imagine what had been haunting the person who wrote it.

"It sounds like a riddle," commented Ishmael.

"Perhaps," ventured Atollana, "though I suspect, rather, that these were just *feelings*. Yes, for myself I see these as simple feelings which have not been translated into technical language. Language can be such a rigid messenger, forcing us to fit our thoughts into whatever words others have already invented. We are assuming her poem has an answer. But maybe

---

[7] *Honua* is a Hawaiian word meaning earth, world, or ground — ED.

it has neither answers nor non-answers. And we are also assuming, of course, that she believed someone, someday, would read it."

Ishmael tried to push on beyond the unanswerable. After reading through the poem yet again, he pointed to the first line of it. "Are we spinning?" he asked, "and what of dew, why does she speak of the dew as though it had a will of its own?"

"It doesn't," Atollana speculated. "At least I don't think that it has. Dew never seems to come and go as it chooses. It always tends to follow the elements. Of course, maybe that's precisely what it wishes to do."

The Dew

Neither shrimp spoke again for a while. They sat on their cozy coral couch and gazed up at the moon's light filtering through the water. After a while, the rays of light began to dance on them because of the refracting effect of dew which had come to settle upon the sea. Atollana studied the dew flirting with the moonlight on the ocean surface, and this gave her an idea. Turning to Ishmael, she broke their silence :

"Yes, maybe that is what the dew *wants* to do, Ishmael. Tell me, is there some element, some fiber or fabric of the Universe, which forms a link between the sea, the terra firma, and the sky? Something which never speaks, something which sees all, something which appears immune to time?"

Ishmael, flustered, looked back at the shell and asked her to explain her philosophical rambling.

"Ishmael," she insisted, "I believe that this is what our unknown friend was speaking of."

"Who?"

"This *Honua*. The terra firmate woman who wrote the poem on the shell. Ishmael, the dew! Remember the dew? Look above you at the roof of our world. Do you not see the dew resting on the water? The dew kisses the sea, it caresses the terra firma, it embraces the air. The dew is a fabric of the Universe. If we follow it, perhaps it will show us from whence it spins."

Atollana put a hand over her chin and mouth to help her think, surveyed the surface of the water again, and then turned once more to Ishmael.

"It's as if," she continued, "the girl might have said, 'Find the point from which we spin, *for there the dew and I begin'*. Come, Ishmael, let's follow the dew. Maybe it can show us what it has seen."

The moon shone brighter and clearer as they rose toward the roof of the sea. Piercing its surface, they saw the dew which had settled upon the water to rest, the moon dazzling brilliantly on it. In the distance they could see the rim of the terra firma, its inland mountains rising amorphously on the horizon. Neither shrimp had ever seen the dry earth so grandly, and both remained motionless in awe of its exotic beauty. When a multitude of black fluttering specks glided harmoniously against the gray silhouette of the mountains, the shrimp were paralyzed with wonder.

"My dear Ishmael," uttered Atollana softly, "I do believe we are witnessing a school of bird-folk skimming through the ocean of air! What a great revelation it must be to view the world from such a distance! How trifling our woes must seem to those lofty creatures!"

Exhilarated, she lost all fear. "Come, hold tightly onto the dew with me. Soon it will evaporate into the invisible sea of air, and when it does, we will evaporate and ascend with it." Pulling closer to Ishmael's ear, she whispered passionately, "we will ascend with it, *the dew and we as one*."

Both shrimp held fast onto the dew, Atollana fixed on the sky, and Ishmael contemplating Atollana's last words. After a very long moment of complete peace, the air gently began to assimilate them into its realm; the dew was evaporating, and the two shrimp were rising with it. They stayed suspended just above the water, then began to rise, very slowly rise. Gradually they gained speed, quietly witnessing the Earth falling ever further beneath them. Their little reef melted into the coastal mosaic fresco, and the Aquanesian Sea became an indistinguishable speck in the furthermost crevice of Earth. Higher and higher they climbed, leaving behind all the seas, the terra firmas, and even, it seemed, the sky. The whole of the Great Macrocosm enveloped them, and the Earth became barely distinguishable from all the other astral organs.

III
Tea - Time

So the two spaceshrimp, marooned in dew, sped on into nowhere. Finding it was quite chilly in deep space, they huddled together to keep warm. The dew was feeling a bit shivery as well, and so it huddled together with the shrimp. It was also rather sleepy in deep space, and so the dew fell into a deep slumber. Then Ishmael and Atollana fell asleep with the dew. But just before they did, a marvelous figure, nestled in Nothingness, became visible in the celestial distance. It was green.

Now the shrimp and the dew had a dream. At first their dream had no picture, no image. Rather, it had only song. And the song had no words, only music. The music was a single voice, the sweet, pure song of a child. But soon the music began to glow, and the glow began to form an image. The image was of the child herself, of the ocean, of mountains, and of an exquisite dawn and thundering rain and everything else that made up the world. From the music all these visions were spawned; it was their molecules, their fiber, their very being. And through the energy of the music, these image-visions entered the tangible Universe. Other parts of the song turned a bit less liquid, grew more defined, and became words. The words of this song had no language. Everything was at peace with everything else. Music was the eternal energy from which all things were created and upon which all things breathed. Music was the common thread which linked all life, the inspiration and power through which all creatures made love with the very soul of Creation. That was their dream.

Tea–Time

The dream passed. The shrimp and the dew continued their sleep, all the while approaching ever nearer the strange green figure. Eventually they reached the object, brushing along supple branches of it, and finally came to rest upon it. The dew quickly awoke and set off on its own, being careful not to disturb the shrimp. Soon, though, a voice began to nudge Ishmael and Atollana from their celestial slumber.

"Buon giorno," the voice said.

Neither shrimp stirred.

"Good morning," the voice said again, "I've made some tea. Clove tea. Would you like some?"

Slowly, the two mis-placed crustaceans, groggy and confused, began to awaken. Atollana was the first to sit up. Fixing her eyes on everything above her, she barely noticed that they were on some sort of terra firma.

"Ish . . . ?" she whispered, "Ishmael, come . . . look around you. *Behold Everythingness!"*

Ishmael, prying open his eyes and peering about, found themselves in the very bosom of the Universe, wholly engulfed by the intense stillness of myriad stars. But rather than allow his eyes to feast upon the celestial offerings at hand, he remembered the queer green object they had seen just before falling asleep, and realized that it was this planet that they were now upon. He surveyed it with gawking eyes. It looked like a watercress.

Hoping that perhaps he was still dreaming, Ishmael closed his eyes and rubbed them. But when he reopened them, he still saw a watercress. This clearly being too fantastic to trust to his own senses, he nervously queried Atollana on the matter. Finally and with great effort wrestling her eyes from the cosmos, Atollana scanned the terra firma they now inhabited and, incredulously, concurred with Ishmael's own judgement. In the blackest immersion of deep, perfectly tranquil space, they lay cradled in the soft leaves of what indeed was a beautiful giant watercress. And it was this watercress that was talking to them and offering them tea.

"A bit nippy this morning," the watercress continued, "Excuse me if I begin my tea without you." The watercress reached for a half gourd — it was his tea mug — and poured a steamy liquid into it.

Atollana was visibly jittery. "Er, um, how do you do that, no, I mean, how do you *do* — ? You see, it was the dew, well, what I mean is I *do* hope that you don't mind our crashing here. By incident. I meant *accident*. We were asleep. Forgive us. Oh, my name is Atollana. From the Aquanesian Sea. On Earth. We're from Earth. My friend is Ishmael. That's his name."

"Yandu's mine," the watercress said. "You didn't crash here. The dew set you down quite gently. And I must tell you how delighted I am that you've come to visit with me. Please do make yourselves comfortable. The cloves have steeped long enough. Tea is ready."

The watercress reached around to select two more gourd-mugs. Ishmael pulled himself up and gazed about, while Atollana gathered the courage to cease trembling and ask this Yandu fellow where they were.

"Er, sir, Mr. Watercress, tell me, what place is this that we are . . . are *on?*"

"Please call me Yandu," he began, pouring fresh tea into the gourds for the shrimp. "Call me Yandu. That's my name. 'Watercress' is a botanical term. That's my phylum. Watercress. By the way I know your Earth. I grew up there. Splendid world. A paradise for watercress. For shrimp also, yes? Incidentally, I was born with the name 'Veradu', for *vera*, or truth, and *du* for 'do', that is, *action*, for I had many ambitious projects. But I discarded that name when I was still a sproutling."

The watercress was so excited about his two visitors that he'd forgotten to allow them a moment to acclimate themselves before drenching them with such an involved talk. But Atollana tried to keep up with his ramblings.

Tea–Time

"Why did you do that? It was a nice name, no?" she probed cautiously.

"A nice name it was. But one day we had an eclipse of the sun. I didn't understand what was happening. Creation, it seemed, was in contempt of known and established principles. I had to re-invent cosmology just to accommodate that one event. It humbled my trust in *Veradu*.

"Then matters got worse. In my most grandiose scheme, I decided to grow myself exactly north-to-south, that is, to make myself form part of a longitude. It was fun to think that I would be aligned with the two poles of the planet. So, knowing that free-floating magnetized objects point north, I made my longest, thinnest, most flexible leaf into a compass by having it, over time, accumulate iron by drinking exclusively from a nearby ferrite-rich spring. This leaf would become a permanent, precise marker for my polar alignment. I went to every pain to insure its accuracy, for example providing my compass leaf with perfectly even shade to avoid its being unconsciously influenced by the direction of the sunlight. Once I was confident that the leaf was pointing the way to the pole, I carefully let my body grow strictly along the line it dictated. But after years of developing along my much-touted line, two terrible truths became wholly undeniable. One, that the line I had formed could not be truly north-south; the path of the sun made that brutally evident. Somehow, the north of my magnetized leaf and the north of Earth were not the same. Secondly, and even more devastating, the orientation of my compass twig had itself varied markedly over the years.[8] I was a sorrowful absurdity indeed, carefully grown so straight, irreversibly pointing the chosen direction, with the sun — an immeasurably higher authority than myself! — in unremitting

---

[8] The explanation for this passage is clear. Even if Yandu had known to compensate for magnetic declination (the discrepancy between magnetic north and true north), his venture was doomed because the magnetic pole is not a fixed point, but is continually, slowly, changing. — ED.

The Crustacean Codex

disagreement, my sad compass limb forever changing its ferrite mind. And all this after my roots were already long anchored.

"Then one fine summer day, while I was still bewildered from my magnetic chaos, it hailed. An impossible phenomenon, I would have believed! Pellets of ice sheared through the balmy weather, poking holes in my leaves of folly and in my dauntless innocence. Humbled, I unceremoniously relinquished *vera* for *yin*, the passive power, and with *du* I honored my inspiration from the *dew* rather than 'do'."

Before either shrimp could untangle this saga, the watercress dove into his commentary yet again.

"Thus I became *Yindu*, for henceforth I would observe rather than meddle, digest rather than impose, and yield to the elements, in the same manner as the dew. Finally, I altered one letter to become *Yandu*, acknowledging that like all mortals — and immortals, I suppose — I also share elements of the active power. Indeed, I have been even busier ever since."

A sudden silence ensued. Neither shrimp could camouflage their perfect incomprehension of Mr. Yandu and his wordy ramblings. Fortunately, however, when the watercress saw their hollow gazes, he finally realized that he had jumped right into a most serious tea-time oration without even giving his new friends a chance to gather their wits.

"Now, would you like fresh mint in your tea?" he asked, inviting a less intellectual response, "and also for you, Ishmael? Fresh mint?"

Both shrimp looked at each other and nodded a confused 'yes'. They were both fond of mint, and were both still rather chilly. The tea smelled tantalizingly piquant.

Still trembling a bit, Atollana raised her mug to her lips, hesitating to meditate on the image of the cosmos mirrored in the tea in an attempt to calm herself. Soon the reflected image of the stars and of her own face was steady enough to assure her that her shaking was nearly under control. Never losing her fix on the tea's mirrored image of the firmament, she waited as a

Tea–Time

lone clove slowly sailed from one constellation to the next, from one galaxy to another. *When it finally comes to rest at the rim of the mug*, she told herself, *I will conquer my nerve and query this gracious fellow Earth person.*

Above all, what she wished to ask Yandu was *where are they*. She was reluctant, however, to ask such a question directly, for fear that it might appear brusque. Instead she decided to ask about another matter that concerned her. The tea's wanderer clove now reached and skimmed the mug wall, skirting several nebulae before finally coming to a stop. Keeping her promise to herself, Atollana plunged forward.

"Mr. Yandu, are we not hurting you, to be stepping all over your beautiful stems and leaves?" she asked, partially overcoming her fear.

Pausing only briefly to contemplate this, Yandu replied: "Back on your Earth, do you first inquire this of the jungle before entering it? I am nothing more than a watercress forest which has grown as a planet. I grew with ample space for friends to visit. By the way, 'Mr' is a title. I don't have a title."

Atollana, still uneasy, tried to explain that she was afraid they were intruding, and then asked: "Can you maybe tell us, Yandu, why the dew brought us here — um, though a pleasure it is — ?"

Yandu raised his eyebrows and almost smiled. "The dew did not bring you, *per se*," he explained. "It wasn't the 'taming of the dew' — rather, the dew came here, and you were with it!" He slowly sipped his tea, then continued. "Look at it this way, Atollana — by the way, I assume that 'shrimp' is merely a zoological term — on Earth, does the dew not come to settle upon the watercress in the morning?"

"It does indeed, sir," she answered. Both shrimp looked at each other and felt comfortable agreeing to this supposition.

"Well, likewise it does here. I am a watercress, and every morning the dew comes to settle upon me. A permanent but vagabond settler, I might add. Now, please do make yourselves comfortable, and I would be most flattered if you would tell me your story."

With this the shrimp calmed further and told the watercress about the baffling and unexpected events of the past day. Yandu was intrigued by their late friend La Vecchia ("I regret that she never came for tea," he reflected), and curious about the story of Pliny Plankton's necklace. Its gold chain and earthy shell were especially pretty against Yandu's green stems, but Yandu and the shrimp together could not shed any more light on its message than the shrimp had managed by themselves back on earth. Yandu, however, insisted that he had enjoyed a tea with an interesting fellow named Pliny Plankton a long time ago. Ishmael and Atollana were of course quite excited to hear this, though it was difficult for Yandu to place his visit in terms of earth moons or years.

To judge time whilst disoriented on a giant watercress in deepest space is especially difficult for shrimp. Day and night do not exist on the plant/planet Yandu in the same sense as on Earth, as Yandu received little light from the many stars in the Universe. But every morning, with the day's first yearning for dawn, dew would come to settle upon the great watercress (which, as Yandu had pointed out, is precisely as the dew does on Earth). And the dew would bring with it warm, luscious, titillating light. Upon reaching Yandu, the dew would begin to offer the light which had accompanied it, first releasing a barely perceptible radiance, ever increasing until Yandu had daytime. So there was a daily cycle of sorts, and it did seem that some time passed.

Yandu took quite a liking to his two shrimp guests, and did his best to make them feel welcome. Their favorite endeavor was chatting over tea; the three engaged in a constant tea-time and discussed many tea-time topics. Of all such topics, The Nature of the Universe was their favorite, for all three seemed to have figured out life in varying ways. Atollana had become fascinated with Yandu, and was determined to learn his tale.

"Mr. Yandu, you have life figured out amazingly differently from anyone on our reef." she began. "Are you not lonely, being all by yourself here?"

Yandu laughed. "Yes! Of course I am! On your planet, most people are lonely as well, except that they've chosen to be lonely all together."

"But I am speaking of your feelings!" Atollana pressed on, "Not of philosophy!"

Yandu thought a bit and then tried again. "Atollana, this is an outpost of sorts. To fear that I am lonely here is to object to water because it makes you wet. I live here *because* it is an outpost."

"An outpost from what?" asked Ishmael.

"From *everything*," Yandu answered. "Absolutely everything. *Don't you understand where you are?*"

The two shrimp peered about the entire, now familiar periphery, squeezed and stretched their scales, then looked at Yandu and shook their heads.[9]

"Well," Yandu began, "I, like you, was fixated with the innumerable ways in which people have life figured out. So I longed to establish a neutral frame of reference to observe things from, a ground zero, so to speak. I wanted to relate things to an absolute. That absolute, of course, does not exist. But I reside in the illusion of it."

"We are at ground zero?" asked Ishmael.

"Almost. Rather, we are *straddling* ground zero. True ground zero is actually beyond my far frontier, quite a journey from here. I deliberately pivoted myself just off-center because I wanted to observe the Universe from virtual neutrality, that is, from as close to perfect neutrality as possible without losing perspective. There is no neutrality, there is no objectivity. Except subjectively speaking, of course. Were I to observe the

---

[9] The term 'scales' refers to the shrimp's translucent exoskeleton. — ED.

Universe from true zero, I would lose dimensionality. At least that was my fear. So I watch from just a bit off-center. It is just like your two eyes," he finished, without elaborating.

"Like our two eyes?" puzzled Atollana.

"Well, yes. Your eyes — and your ears as well — are just close enough to each other so that they can perceive a single image, yet far enough apart so that you preserve dimensionality. That is the precise position of my head in relation to ground zero."

Atollana was getting frustrated by this esoteric geography. "Where, then, are we?" she asked.

"We are just outside the Primal Pivot," he explained flatly, "a vestige of earliest Creation. It is the axis from which the Universe whirls, the place most equally distant from absolutely everything. And, for these reasons, it is the place I chose to root myself. Beyond me lies a sea-cradled island orb which is the Middle, that is, the middle of the Primal Pivot, and at its precise midpoint is the spot which people call the Center. It is that place from which the entire Universe is pivoted."

From which the Universe whirls spun through the shrimp's heads as their eyes fidgeted to the poem etched on the necklace around Atollana's neck.

"But you don't watch from there because you were afraid that the Universe would appear monodimensionally?" Atollana asked.

"Correct. But whether my concern was warranted or not, I don't know. Seems my tea-partners all disagreed with me."

Ishmael and Atollana excitedly questioned Yandu about the Primal Pivot. It could be, the shrimp speculated, that some answers — if not the Answer — might lie at the true Middle and the Center. They were, after all, still only at the outskirts, far from these vantage points of perfect, unimpassioned neutrality (or, as Yandu pointed out, what could be considered perfectly impassioned neutrality, depending on one's point of view). But Yandu couldn't tell them much about it, even though it lay just

Tea–Time

beyond his back. He was not, as they had thought, an omniscient creature who witnessed all of creation from a celestial balcony. Yandu, of course, had never pretended to be particularly enlightened; Ishmael and Atollana had only assumed that any watercress who had communed with the hub of eternity for several eons should have life pretty well squared away.

So, despite Yandu's own decision to circumvent the true Middle — or perhaps because of it — the shrimp were inspired to reach it. They wanted to see how the Universe appeared from theoretical neutrality.

"We would like to discover truths," as Atollana explained it.

"Truths?" scrutinized Yandu.

"Truths," corroborated the shrimp.

"Ahhh, you mean *music*?" the watercress probed.

Atollana was surprised by this interpretation, but politely tried to clarify.

"Well, maybe that also, but I meant more like *knowledge*," she corrected.

"Oh, you mean *facts*?" checked Yandu.

"Well, yes, I imagine you could call them 'facts'."

"Quantitative measurements, phylum 'facts'. Now I understand what you mean. So, tell me, what *kinds* of 'facts' are you looking for?"

"Oh, are there different kinds?" queried both shrimp, accidentally in unison. With this response, Yandu realized that a major tea-time was brewing. He clasped his hands and leaned back a bit to fix his mind on the new topic.

"Yes, yes indeed, distinct varieties," he began, "there are many different types of facts. For example, the most revered are True Facts. These are facts which can be empirically demonstrated, and cannot be experimentally disproved. They

The Crustacean Codex

are stable creatures, changing only infrequently. For example, *'light is a constant'* is currently regarded as a true True Fact.

"But True Facts are not necessarily the most useful kind. Some True Facts are utterly useless, as I will demonstrate." Yandu picked out a volume entitled *True Facts* and referred to an appendix at the end. The appendix was headed *True Facts with No Known Use in Chamber Music*.

"Ah, yes, here's a good example. I will read it to you: *'Glass is a liquid at normal Earth temperatures'*. I can't tell you how many times I've played sonatas without ever having found a use for that particular fact. But perhaps physicists know what to do with it, if only I could get one here for tea. Certainly, True Facts have earned the respect they receive. Please let me know if you learn the use of any facts in this Appendix, so I can correct my records. And I am most anxious to add to my compendium of facts of any sort, should you ever happen upon any."

"I have a new fact to offer," ventured Atollana, "it is an absolutely True Fact with a bright future of very many uses."

"Excellent!" jumped Yandu. "I will add it to my compilation of True Facts. I have my notebook handy. Please tell me!"

"Here it is: *'One plus one equals two'*, or, perhaps better put, *'a' plus 'b' equals the total of 'a plus b'*. That is a True Fact if ever I've seen one!"

Contemplating this seemingly innocuous statement, Yandu held his quill in the air above his notebook. But in the end he put his hand at ease rather than record the exciting new fact.

"It is an interesting observation you have made," he finally said, "and it is indeed a worthy fact. But I cannot list it in my book of *True* Facts. Instead, it will fit nicely into another book I keep."

With that, Yandu reached behind him and picked out a notebook entitled *Observations which Function as True Facts When Applied to Grains of Rice*. Seeing that Atollana was focusing on this lengthy title, Yandu offered a simplification.

Tea–Time

"I call them 'Rice Facts' for short. One grain of rice plus one grain of rice equals two grains of rice. The formula you offer is promising as a Rice Fact, but of dubious value beyond that."

"Oh, but I must disagree!" protested Atollana. "It is an eminently useful, fine, full-time, absolutely *true* True and very handsome, debonair fact. One plus one *always* equals two. A more charming, delectable fact never will you meet!"

"Doubtfully," insisted Yandu, still digesting the implications of such a fact. "I fear it is a flawed fact. For example, if two people make music together, each person's sound with a loudness of '1', it would *not* result in sound with the loudness of '2'. Normally, it would yield a sound with a loudness of *less* than '2', and under very special circumstances it could even produce a sound of *less than '1'*. Certainly this helps explain why fine chamber musicians are so neurotic. And why they tend to cook too much rice when making dinner."

"Well, you remind me," commiserated Atollana, "one evening some coral had a chorale party, and the left-over food fed the entire reef for a quarter-moon."

"Indeed, seems they could have used your new fact," Yandu continued, "as long as they remembered to compute water absorption into the equation, of course. Yes, as a 'Rice Fact' I find your suggestion most appealing, and with your permission I will enter it here in my log. I suspect the critter will prove very useful for preparing feasts, as well as for other important ventures."

Ishmael inquired about facts which perhaps were *once* True Facts, but which were then demoted to other ranks of facts. There was apparently an impressive inventory of them from which to choose, they learned from Yandu.

"Formerly true facts can be very interesting. I catalogue the archaic ones," the methodical watercress explained, handing the shrimp a notebook entitled 'Antique Facts'. They opened it to a random page and read an entry which said *'the sun rotates around the Earth'*.

"Oh, you picked a favorite of mine!" exclaimed Yandu. "Has a lovely patina to it. That entire notebook contains antiquarian facts. They were once, reasonably and understandably, considered true, but are now known to be untrue.

"On the other extreme, there are 'False Facts', observations which are currently classified as 'facts' even though we know them to be untrue, and which, whether for reasons of convenience or poetry, we keep handy. But a False Fact is not necessarily a *bad* fact; there is a special breed of them which is actually very positive. Since people are most fond of these specialized facts, and they receive much petting as a result, I call them 'Furry Facts'.

Yandu pulled out a book labelled *Furry Facts* and skimmed through it briefly. "Here are a couple of nice examples. '*All things are possible*' and '*all people are equally capable*' are wonderful facts when used properly, even if they are not really true."

Turning to the next page, he noted another entry. "'*Things fall because gravity pulls them down*'. How about that one? Oh, here's another good one: '*An elected leader is the choice of the people*'. A *very* furry Furry Fact, indeed.

"But while Furry Facts are usually magnanimous creatures, their alter-egos, the 'Chameleon Facts' and 'Fickle Facts', are often sinister. These slippery beasts constantly adapt to new surroundings or to the whims of their masters.

"For myself, I have found it useful to categorize all the various fact factions in this way, and I keep a separate notebook for each. But on Earth, facts are organized in another manner. As I'm sure you know, on Earth, various people are responsible for institutionalizing different species of facts. It is an interesting phenomenon."

The shrimp glanced blankly at each other, and Ishmael spoke for them both.

"No, we've never thought about it. How does that work?" he asked.

"Oh, well, for example, the scientists are responsible for codifying *True* Facts, by which I include such variants as Rice Facts. I really admire those fastidious factitioners."

"But what about the False Facts?" questioned Ishmael.

"False Facts? The clergy, of course," answered Yandu. "They gather faithful flocks of False Facts. They have a fetish for the poor creatures. Indeed, False Facts have been known to expire from exhaustion under the clergy's tutelage, so unrelenting is the friar's mission with those of their fold. And Romanticists groom Furry Facts, while politicians breed the Chameleon Facts and Fickle Facts."

"And the many facts in between?" pressed Atollana. "What about them? All the vast world of fuzzy facts? Have they no home on Earth?"

"Not to worry," assured Yandu. "They have quite a loving home indeed. Those endearing mortals are tended to by philosophers and poets. Admittedly, however, they often live a restless life. I've known many such facts to become extremely self-conscious around philosophers, ultimately losing all self-esteem. These unwitting victims of the philosopher's scalpel would be relegated to a life of despair were it not for the poet, under whose brush they almost always find solace."

"So once these facts enter the protection of a poet, they live a peaceful life?" asked Ishmael anxiously.

"Well, not necessarily," corrected Yandu. "From there they may eventually come under philosophical scrutiny once again, leading them to seek for renewed poetic salvation."

"It must be a truly difficult life, being a fact," reflected Atollana gloomily.

"Not an easy life at all," concurred Yandu. "People expect so much from them. And facts are, after all, only human. They are vulnerable to the same temptations of the flesh as you or I. At any time a Furry Fact might succumb to excessive petting and become a False Fact. Or a False Fact might dress in quantum clothing and be mistaken for a Rice Fact. We burden facts with such responsibility. Sometimes we need to be more patient with them.

"By the way," Yandu continued, "there is another species of fact, a very precious one, which we must *all* care for. If something evil happens — if an oppressor inflicts misery and suffering upon people — these facts are very sacred. Certain people may wish to deny them, to erase their memory, to create Unfacts or Antifacts in their place. We must always protect the Truth against them.

"I would be very interested to know the nature and caliber of any facts you encounter in the Middle, and how those facts fare in that world. And do especially let me know if you find any Antique Facts. It is important that I record them before they are lost to posterity."

"But won't you come with us?" asked Ishmael.

"I would very much like you to go to the Middle," the watercress replied. "But I cannot go with you. I must remain here. I am rooted here."

"But why?" Atollana insisted. "You can leave your roots here for the time. On Earth, all watercress have both literal and figurative roots. They keep the figurative roots while taking liberties with the literal ones. This must be some kind of a fact, I imagine. Our roots are in the Aquanesian Sea. But we left them there for now. Please do come with us."

"My dear Atollana," Yandu explained, "don't you understand how impossible it would be for me to go hiking with you along my back to a place which lies somewhere beyond it? This reminds me of the legend of that able terra firmate named Atlas."

"I don't know the legend," Atollana confessed.

"Well," he explained, "some irate deities obliged the poor fellow to support the earth and sky upon his back for eternity. Imagine how heavy the sky must have been on those overcast mythological days, when all sorts of bizarre things went on in the heavens! Could you imagine Atlas wishing to stroll through the hills and woods? The Center, of course, does not rest on my back — but my final, thin limb just meets the ocean sea which encircles it, and I don't want to disturb this alignment. You two should certainly go on an expedition to the Middle, however. It's very pretty there, or so I've been told. I can't see it from here. In fact, I am told that one cannot actually see it from without. It is a phenomenon which I'm afraid I cannot explain."

IV
Journey To The Middle

Such mystery further motivated Atollana and Ishmael to undertake a journey to the Middle, the Primal Pivot. While stressing that it would be an arduous hike, Yandu was encouraging to the shrimp, giving them detailed directions and joking about how he would love to go with them if only he could find someone to shoulder the cosmos during his absence. The shrimp, of course, knew that he was teasing, that he was not an Atlas-type kingpin for the Universe, but rather just a cosmically insignificant observer who had chosen to reside next to the Primal Pivot.

Yandu advised that they follow one of the paths which runs along his mountainous spine. They were all about the same length and difficulty, but he asked if they would follow a very specific route as a favor to him. "My back is so sore," he explained, "it desperately needs a massage. If only I could get someone to walk on the right part, I would feel much better. You see, once a year my entire body is kneaded by cosmic tremors, but it has been nearly a year now and I am quite stiff. I would be most grateful if you could go by way of my back mountains."

The two shrimp were, of course, glad to oblige. They asked what these 'cosmic tremors' were about, but Yandu could only tell them that every year when his body was aligned in a certain direction, it vibrated (and even shook) for a period of about a day, the disruption then subsiding and everything returning to normal. It was very therapeutic, if inexplicable.

The tremor discussion was their last tea-time together. The day had come for Atollana and Ishmael to part company with their friend Yandu — or, rather, to part company with the part of him that included his head and arms, and set their course through his mountain forest and on to the Primal Pivot. They marched off with Yandu's directions fresh in their minds just as dawn began to spread its glow on the watercress mountains in the distance.

Yandu had pointed out a high, thick forest that lay upland and was separated from them by a misty valley. They were to veer to the left when descending into this valley to avoid a particularly steep bank. To their surprise, there was no path to reassure them that they were on the proper course, but then nothing in the forest seemed particularly menacing, and their immediate destination, the Yanduian Highlands, was almost always in sight.

When they reached the lowest point in the valley, a misty steam which constantly filtered through the valley floor clouded their footing. The steam itself was not very hot, rather just warm and wet, and the footing, they soon realized, was not treacherous; both shrimp felt relieved, however, when the ground finally sloped upward toward the forest atop Yandu's mountainous spine.

Once within the forest cover they found a path which had been worn into a ledge near the top of the valley. Although the valley wove wildly about the entire narrow mountain range, they were pleased to discover that the path itself avoided the mountains' endless undulations by switching mountainsides at opportune passes.

The valley forest was a friendly forest. The shrimp, however, found themselves unnerved by it, for it seemed to be watching them, or at least, they thought, to be *noticing* them. They walked without speaking, peering carefully about out of both trepidation and wonder. Finally, Ishmael tried to joke.

"Shrimp shiatsu," he commented. Atollana was very quiet, and stared at the weird terrain around them.

Ishmael tried again: "Shrimp shiatsu, Atollana, that's our mission up here on Yandu's spine. He must be enjoying this. Remember, he said that it has been nearly a full year since those tremors massaged his spine."

Atollana, though thoroughly charmed by the mountain's garden forest, pulled herself from its grip. She tried to joke back.

"But Ishmael, of what use will our feeble massage be to Yandu? He does not have enough mass to create much planetary gravity, and we are too light to exert much pressure on his sore back."

Ishmael thought about this. He began to explain to her that they were neither capable of modifying their weight nor Yandu's gravity, and that this was indeed an exceedingly true fact. But before he finished, Atollana, who was walking behind him and found the discourse thoroughly amusing, leaped into the air and landed squarely on his back. Ishmael jumped in fright.

"It is only I," Atollana assured him, laughing. "You are wrong about it, you know. Yandu would have classified that as only a 'foggy fact', I think. You see, both of us walking on one set of legs will double our weight and massage Yandu's back twice as well!"

So the two of them hiked piggyback over Yandu's long mountainous spine, occasionally deviating slightly from the path but always staying within the central valley and its upper rim. They would switch positions when the one on the bottom became tired, stopping to rest when they both grew weary. Ishmael at one point suggested that the one underneath should walk on as few legs as possible to further concentrate their weight, but that proved to be impractical for shrimp. Perhaps La Vecchia had been correct; perhaps there were some advantages to terra firmate physiology.

All along, their forest had been changing, subtly becoming more primitive and dense. It was this that had begun to preoccupy only Atollana earlier, but by now both shrimp were succumbing to its raw beauty. They no longer walked piggyback. They no longer stopped to rest. And as the forest had grown too thick to walk side-by-side, Ishmael walked ahead, hoping to escape its evolution to the fantastic, while Atollana walked behind, hoping to capture it. Neither of them spoke, as if unwilling to disturb the forest Being.

Suddenly, turning a bend, they reached the end of the forest, and the end of Yandu proper. They confronted a sheer precipice out of which there grew an immense, winding stretch of branch, slightly flattened on top. At the end of it, a distance difficult to gauge, there rose the figure of a mountain, albeit an unseen mountain, swathed in a silvery mist. The bottom of it was circumscribed by what appeared to be a golden bracelet. Altogether, it had the imagery of a primeval volcanic island shrouded in a silver cloud and encircled by a golden sea. Both shrimp reckoned that they beheld the Middle of the Primal Pivot, their destination. And both shrimp realized that the long branch leading to it was the 'final, thin limb' of which Yandu had spoken. Never had they expected it to be so long or stark.

Atollana expressed concern about the narrowness of the branch and the worry that if they slipped they might fall off into space. Ishmael wondered if instead gravity always regarded 'down' as whatever direction was toward the *terra* (in this case being Yandu's branch) but certainly did not wish to risk being wrong about this. They would be careful to stay on the 'top' side of the branch.

Ferdinand and Guinness

Although neither Ishmael nor Atollana had expected the hike over Yandu's back to be so long or so odd, the prospect of a journey on this final, long precarious twig of Yandu made the woodlands seem benign, even comforting. For as queer as the forest had been, it had mercifully let them forget that they were alone. When they reach the Middle of the Primal Pivot, they would confront perfect solitude and attain profound and absolute obscurity. Such were their meditations as they beheld the outpost of Nothingness, Yandu's so-called 'ground zero'. But, suddenly, those thoughts were interrupted by a frantic rustling in the forest just behind them.

Two small and dark four-legged figures skittered from the forest edge. The mysterious creatures rushed up to the two shrimp in a great frenzied excitement.

"What's going on here!?" demanded the closer of the two figures, in a low, gruff voice. The other looked on zealously as he jumped up and down and wagged his tail. Ishmael and Atollana were paralyzed with terror.

"What's going on here," he repeated, "and to where do you wish to be herded?"

As Ishmael and Atollana tried to collect their voices, the creature that had spoken looked back to the other for approval. The latter wagged his tail faster in admiration.

Atollana tried with all her strength to calm herself and to respond intelligibly. Shaking uncontrollably with fear, she cautiously uttered, "Good day, sir, ahh, sirs, we are two shrimp passing through these woods at the invitation of Mr. Yandu. We did not mean to trespass. I am Atollana and my friend is Ishmael. I do beg your forgiveness for our intrusion . . . and for any trouble we may have caused. Please do not harm us!"

For a few moments there was no response to her desperate introduction. The dark figure tilted his head and oscillated glances between his enthusiastic accomplice, who was still fretting about near the forest thicket, and Atollana.

"Hmmm, a stranger in a strange jargon . . . have you passports?" he inquired of Atollana.

Atollana was flustered by this interrogation but confessed that they certainly did not.

"No passports? Neither of you? Blimey!" He now turned around and spoke to his little friend. "Guinness, go get these two travellers passports. And remember to bring the official Quake Sign, in case they've never read it, as well as anything else they might need on their herding."

Atollana tried to speak again. "Excuse me, sir, are you a four-legged terra firmate?"

"Well, if that's Latin for 'sheep dog', then you are correct. Just call me Ferdinand. Ferdinand is my name. And my apprentice is Guinness. That's his name. Guinness." With that, Guinness, number two sheep dog, rummaged back into the woods.

Although Atollana had never heard of sheep dogs, she was familiar enough with the writings of Pliny Plankton the Elder to know that 'sheep' and 'dogs' were both varieties of four-legged terra firmates. Perhaps Pliny had known about sheep dogs; but quickly her mind returned to the alarming situation at hand.

"Pardon our surprise, sire, um, sir, Mr. Ferdinand. We've never met a sheep dog before. We've never met any terra firmate before."

"Never met a sheep dog?! Preposterous! My dear child, you have traversed the reckless path of life without proper herding? Has fate left you to plodder through such tender years without a sheep dog to herd you? Oh, how merciless! Then all the more urgently, I inquire of you again: to where do you wish to be herded?"

"You really want to herd us somewhere?" trembled Ishmael, unsure of what might be involved with being herded.

"Want? It is not a question of desire or need," answered Ferdinand, "but rather of the grandest of designs :

I am a sheep dog and you are herdlings."

"But we are shrimplings," Ishmael politely objected.

"Regardless and irregardless. You are still herdlings."

"Well, sir, um, ah," stuttered Atollana, "this is all very kind of you, we really should be able to manage on our own, I think. We are bound for the Middle."

The discussion was then interrupted by commotion in the forest. Guinness, the apprentice sheep dog, returned with some odd junk clasped in his jaw. He distributed two small tattered booklets to Ishmael and Atollana, then brought the rest of the junk to Ferdinand. Among the items were a very old wooden sign, a piece of half-rotted parchment, and some peculiar paraphernalia that reminded Ishmael and Atollana of debris they had encountered lying on the bottom of the Aquanesian Sea back on Earth. Ferdinand now announced their new assignment to his pupil.

"Guinness, these two weary travellers have conscripted our services. We are to herd them to the Middle, being the coordinates of zero degrees on all 'tudes."

Guinness, excited by this news, sat himself down to the side and just behind Ferdinand, a position which the shrimp figured must somehow be pre-ordained by the Natural Order of Things Unknown. The little apprentice dog was glowing with happiness to be embarking upon a new and exciting expedition, and gushing with pride to have so quickly procured the essential prerequisites for it. Meanwhile Ferdinand, number one sheep dog, was trying to calm the shrimp, who were still quite frazzled. Perhaps, he thought, a relaxing brew was in order.

Ferdinand dipped four hollowed Yanduian coconut shells into a small frothy well and passed two of the coconut mugs on to Ishmael and Atollana. He motioned them to drink. Hesitantly at first, the shrimp carefully sipped a warm liquid from the mugs. It was a strong, quite earthy beer, which both shrimp adored. They began to calm a bit, and after some conversation with the sheep dogs, they even began to tease. Atollana wanted to name the beer after Ferdinand, a leader who on this moon shined with valor, while Ishmael thought the moonshine should be named after Guinness, who was, admittedly, a rather dark and stout fellow.

While the four were enjoying their libation, Ferdinand tried to teach the two shrimp to pronounce his name properly (it is apparently a difficult sound for shrimp). Ferdinand got a bit frustrated, but both Ishmael and Atollana soon mastered it. The shrimp practiced enunciating his name by asking him many questions about their impending herding, despite Ferdinand's own suggestion that they relax for the moment.

The imbibers imbibed and the shrimp's apprehension eased, it was time for Ferdinand to present a discussion of the journey to the Center, which would be his Official Herding Briefing. Ferdinand, it was clear, took his life's calling very seriously. He began his presentation by clearing his throat.

"*Humpfff*, so, let us begin our undertaking. It is my responsibility to brief you on the dangers of our expedition, and to prepare you for a safe and comfortable herding." This introduction confirmed the shrimp's impression that Ferdinand was a very responsible sheep dog.

"*Ahem*, now, Guinness has already given you passports," he continued, "which will see you through any diplomatic tangles that might arise. But I must now speak of quite another matter: there are numerous geological faults along the narrow branch that we must traverse. It is imperative that you understand the tectonic difficulties which face us on our herding, so I will ask Guinness to display the Quake Sign."

The number two sheep dog, following this cue and fizzling with anticipation, held up the battered wooden sign he had fetched from the forest. Ishmael and Atollana looked at it and wondered if perhaps he had brought the wrong one. It said :

"Please Do Not Tickle Me. I Might Sneeze."

"Sneeze? Who might sneeze?" Ishmael asked timidly.

"Well, *Yandu*, naturally," Ferdinand answered in an authoritative but reassuring manner. "His sneezing could be catastrophic. It would be geologically Richter. Ours would be just a sneeze."

"But why would we tickle him?" Atollana asked.

By this point Guinness was jumping up and down with such fervor that the wooden sign began to fall out of his jaw. Ferdinand rolled his eyes in exasperation, but then turned back to the shrimp and tried to explain the sign's tectonic message. It was an important message. It seemed that the long, thin winding branch of Yandu that they were to traverse was quite sensitive. Travellers had best take care not to tickle the limb with reckless walking, lest they create a tremor and be thrown off the narrow footing and into space.

"Quake danger, you realize, especially with twenty-eight legs negotiating the branch. But only during this brief quake season. Actually, we seem to get quakes about this time of year regardless of what we do. Don't know why. They come for about a day and then disappear. Now, please examine the map that Guinness has brought you."

Neither shrimp fathomed what map he was referring to, so Guinness, a bit calmer now, wagged on over and pawed out the partially-decayed parchment scroll he had retrieved from the forest. Ferdinand continued his lecture :

"At the end of the branch there is a moat, a circular sea, so to speak, which surrounds the Middle." Guinness pointed the moat out to the shrimp. It was the golden crystalline halo at the end of the branch which they had observed earlier, presumably the ocean sea mentioned by Yandu.

"You must swim through it to get to the Middle. Once we finish herding you to this moat, you will be on your own. The chart will help you find your way once you are within the Middle."

Atollana carefully unrolled the chart, and attempted to read its title aloud :

"Yandunius Centralissimo, Quod Vulgo
Cavum Ater. Pliny Planktonius, Auctore."

She nervously glanced at Ishmael and the sheep dogs. "Pliny? You mean Pliny Plankton? Is this Sir Pliny the Elder's chart?"

"A wonderful and curious old chap," Ferdinand reflected. "I got quite an education herding him. But that inscription means nothing. It's what we call Piglet Latin. That means we can herd it and forget it. And I herded[10] it many times myself, to be sure. There is also a Pig Latin, but Pig Latin is altogether unmanageable. They eat roots. Always nosing for roots. Greek roots. *Roots Homericus*. Roman roots. *Roots Virgilicus*. Such a bother when you're worried about 'quakes in the vernacular."

Ishmael immediately thought of Yandu, whose well-anchored roots had precluded his joining their expedition, and wondered if Latin roots, similarly, kept *words* from straying too far. He knew that 'pigs' were a kind of four-legged terra firmate, and hoped that perhaps some sea folk had their own Latin as well. "Is there a Shrimp Latin?" he asked.

Guinness, who had been panting excitedly, suddenly got quiet and let his eyes and ears droop a bit. Guinness knew that the answer to Ishmael's question would be a sobering matter. He lowered his face and looked to Ferdinand, who spoke in a very learned tone.

"Tradition," Ferdinand began, "records a manuscript, which we must consider apocryphal, entitled *Codex Chroniculum Crustacea*. It may or may not be extant. But that is the only reported example of Shrimp Latin."

Atollana then noticed another inscription on the chart. This one was in the lower part, and was encased in an elegant decoration. It was surprisingly intelligible. "Listen to this," she said, "there's more down on the bottom. I'll try to make it out:

'This Hvmble Mappe Be of Ye Middvl of Yanddv,
Region Compriseth the Centre of Everythingness,
and Domain Which Hath the Vision from With'n
Forever Trapp'd by the Raptvres of its Soul'."

She slowly pulled her eyes from the chart and turned to Ferdinand for a clue to the meaning of the strange words.

---

[10] Original manuscript = 'hearded'. — ED.

"All that means," Ferdinand explained, "is that light which enters the Middle becomes lost. That is why we cannot see the Middle from here." Ferdinand walked to the edge of the small plateau they stood on and looked out over Yandu's long, narrow branch toward the Primal Pivot.

"When we look at the Middle," he continued, "we see only a confusion of dizzy light. Once you are inside the Middle of the Primal Pivot, you will not be able to see back onto Yandu either. Even though nothing will be blocking your vision. That's what the inscription on the chart tells us. As far as I can figure it, light simply has little motivation to leave that place. But we have never been there and do not understand it."

Ferdinand stayed by the ledge, looking out onto the precarious and immensely long branch. Far in the distance was the imposing silhouette of the Center, surrounded by its golden sea-moat. Guinness, Ishmael, and Atollana walked over and joined him. He spoke more of their journey, without moving his eyes from their fix on the branch.

"When we reach the moat, there will be one final task for us before we relinquish our herding responsibilities. That will be to instruct you on the proper use of the swimming equipment that Guinness procured for you from the jungle."

Guinness turned and admired the contraptions that Ferdinand was referring to. They were the things that the shrimp had found somewhat familiar, having on occasion seen them abandoned by terra firmates in the Aquanesian Sea.

"They will be kept in our supply pack," Ferdinand added, "and when we reach the moat, we will set them up for you."

That seemed to finish Ferdinand's Herding Briefing. After a long, introspective pause gazing out over Yandu's winding branch, Ferdinand went into the forest again. The shrimp thought they heard him talking to someone while he was still out of sight in the woods.

Shortly Ferdinand returned with the four coconut mugs, now filled with a steamy purple beverage, and passed three of the mugs around to Guinness and the shrimp.

"Sir Ferdinand," exclaimed Atollana, "this tea is sumptuous! Thank you so much! But how did you get it so quickly?"

"Yanduian springs," Ferdinand explained. "There are hot springs in the forest which bathe the roots of fragrant Taprobana plants and jungle herbs.[11] When the liquid spouts from the ground it is already as you have it now. Each spring is slightly different in its mixture. There are also springs which prefer to send their water directly to the surface, clear and unscented. I can fetch you some of that *aqua pura* to try if you like.

"Before undertaking our herding," the master sheep dog continued as he watched Guinness and the two shrimp sip their Taprobana tea, "we should have a goodly meal. I have just spoken with friends of ours in the forest who are now snuffing out a robust feast for us. We'll eat very soon."

"Oh, dear, that's so kind!" Atollana politely interrupted. "Can we go help them?"

Ferdinand and Guinness exchanged glances and smiled. "No, no, they are happy to prepare dinner for you. Gathering your meal is their profession. It is their obsession. It is their art. We will pass on your thanks to them of course. They are reclusive and prefer not to leave the forest."

Ferdinand bade them rest and enjoy their tea, mistakenly supposing that the matter of helping with dinner was settled.

"But why," queried Ishmael, "should these charitable people whom we've never met go through such a fuss for us?"

---

[11] Taprobana probably refers to Ceylon (for which Taprobana was an ancient name), or possibly to Sumatra (to which it was sometimes mistakenly transposed). — ED.

"Because, as I told you," Ferdinand iterated, feeling he was now being redundant, "that is what they do. That's what they breathe. It is their music."

When the shrimp still looked bewildered, Ferdinand tried yet again, his vocal chords now noticeably tense : "Because . . . because . . . Because they are truffle hounds. That's the reason. They are truffle hounds . . .

They are truffle hounds and you are hungry!"

"Truffle hounds?" asked Atollana.

"Yes, of course. *Truffle hounds*," verified Ferdinand, thoroughly composed once again.

Atollana thought carefully. 'Hound' definitely sounded to her like a four-legged terra firmate, but 'truffle' did not. With considerable help from Ferdinand, the shrimp soon understood that 'hound' was indeed a four-legged terra firmate, in fact a close tribes-person to 'dog'. Whereas dogs generally *led*, hounds generally *searched*. They were complementary callings.

'Truffles' were more difficult to explain. Ferdinand said that they are a plant, and that they have a most special and poetic mission.[12] It was the job of the truffles to save the living germ from dying trees and pass it on to other life. Through the truffles, the life energy of these trees never really perished, but instead was saved for weary travellers such as the sheep dogs and the shrimp. Thus trees could die content, knowing that the truffles would not let their vital cell of Creation perish with them.

And 'truffle hounds', then, were something like sheep dogs who devoted themselves to the herding of truffles. Without them, the germ of life which the truffles saved from the dying trees would not reach the other life which needed it.

All this finally sorted out in the shrimp's minds, it was already time for eating. The sheep dogs disappeared momentarily into

---

[12] Truffles are a fungus (actually a fungal fruiting body), not a plant. — ED.

the edge of the forest to retrieve the feast from the elusive truffle hounds. The meal was absolutely luscious.

The quartet began their journey immediately after they finished. Ferdinand took the lead, carefully stepping onto the narrow branch from their little plateau. He hoped to set a fine example for the others. Then Atollana stepped onto the branch, followed by Ishmael and, lastly, Guinness. It would be Guinness' responsibility to guard the rear of the herd and to continually monitor the two shrimp's welfare. The shrimp were relieved to find that Yandu's branch was not nearly as precarious as they had feared; it was sturdy, did not shake, and was generally wide enough to allow for an occasional mis-step.

So the little caravan began its long journey to the ocean sea which surrounded the Middle of the Primal Pivot. At first, each of the four retained their individual eccentricities, Ferdinand poised as a confident and kind leader, Atollana intrigued and intense, Ishmael looking overwhelmed but doing fine, and Guinness wired to a frenzy. But soon all four mellowed, and they walked calmly for a long, long way.

There were many turns and bends along Yandu's long arm to the Primal Pivot. Some of them were difficult because they were very narrow, mountainous, or both. But others were pleasant elbows which provided comfortable resting points. After covering a considerable distance they came upon one particularly lovely spot, and Ferdinand suggested that they set up camp and sleep. No one objected.

During their sleep, dew drifted from afar and settled on the branch. All four had a dream, and their dreams were all similar. They each dreamt of an empty universe, a universe which bore no stars, no planets, no sheep dogs, *almost* nothing. In their dream, there was one lone creature who comprised all of Creation. That solitary life was a two-legged terra firmate, a simple looking human girl. Unaware that she was Conception itself, the girl wandered the universe Void in search of other

The Crustacean Codex

voices to the counterpoint of Creation. But there were none, no flowers, no people, no wind, no *terra*. There was no polyphony, there was no fugue.

Intent, the girl began to paint a universe.

She sculpted radiating stars and graceful comets, tree-capped forests and thriving oceans, fog-draped lakes with water lilies, mountain-pierced clouds half cloaking a huge golden moon. She brushed warm shimmers onto the stars and seeded secret stalactites in the rock, adding wild winds to the peaks above and all sorts of wonderful quirks to the lands below. She created every creation for which language had, or might never have, a word. And thus she undid Nothingness.

Finally, in a moment of impassioned inspiration, she painted a terra firmate woman, a variation of herself, and set her down upon a terra firma. The home she chose for this sister was her finest *terra*, her most exquisite orb.

In painting Creation, the girl had worked from the periphery of Void-dom and inward. She continually moved back from her work, never actually stepping foot within it. When she reached the middle, which is to say the center of Creation, the pivot of Everything, she stopped short of finishing the entire Universe. Rather, she left the very last bit as it always had been, unpainted, uncreated. All that she had begot sprawled about her except for this small circular place, just wide enough for her to comfortably walk about in.

But she was now corralled in the middle (or, to use a terra firmate phrase, 'backed into a corner'), for the little grotto, this lone rift in the fabric of Creation, was the sole remaining unpainted cove in which she could hide.

The girl remained in her peaceful refuge for a long time. Eventually, the world outside began to beckon her to join it, to abandon her little uncreated den and step into the Universe she had sculpted. After contemplating this at length, she gathered all her courage, took a deep breath, and carefully stepped from her hermitage.

The very last place she had painted was, naturally, the first place she stepped into. It was a little island with a towering and vital primeval jungle, surrounded by a moat of sorts, an ocean sea cradling the island like a womb. This island had been her final inspiration, and she loved it dearly, for it reminded her of the planet she had set her beloved sister upon.

She could not see beyond this jungle-island, nor could the Universe beyond see back to the island, because all light within it spiralled around its unpainted center, the girl's little monastic void, as if wanting to paint it in. The light would flirtatiously dance to the charms of the little uncreated circle until becoming thoroughly intoxicated from it, then falling to the ground in absolute and utter exhaustion. After regaining its strength, the light would begin this ritual dance once again, forever seduced by the little unpainted Center, the lone relic of eternity which it could never enter.

Although the girl made the dew-island her home, she resolved that the old void in the middle would never be painted in. She had grown to regard it as a sacred and comforting place, a refuge of perfect peace, her monastery; by leaving it unpainted, she was defying the ultimate temptation.

It was a little whisper of innocence forever spared the burden of Creation.

The Moat of Dew

The dream faded from the four nomads.

Guinness was the first to awake. He quietly crept up to the highest point on the branch elbow upon which they were camping, being careful not to disturb the others. From atop the hilly mound he surveyed the branch in both directions as far as he could see, wanting to check that all was well before waking the herd. He performed this exercise constantly when they were hiking; as number two sheep dog, it was his responsibility to run up all the little hills to scan the terrain, then quickly scurry back to guard the rear of the procession.

The day's first glow of dew-light gave Guinness only a silhouetted glimpse of the branch in either direction. He had, however, been sharply trained by Ferdinand to extract as much data as possible from such a limited view, and was confident that all was well. The number two sheep dog then set about to prepare hot tea. Although there were no Taprobana springs along the branch, he did locate a sturdy Molucca bush which had grown in a nearby sandy patch during their sleep. With a few of its leaves and the water from a clear hot spring, he improvised four mugs of hardy tea. The Molucca aroma gently awoke Ferdinand, Ishmael, and Atollana. They sipped the broth while quietly ingesting the Yanduian dawn, and soon resumed their journey. No one spoke of their dream.

The four kept a strenuous pace, climbing and resting, hiking and sleeping. As Yandu's long branch received little of the light couriered by the luminous dew, the two Earth-shrimp were seldom sure what part of the day it was supposed to be. But there was still a vague dusk to remind them of fatigue, and a subtle dawn to mark the new day's genesis. Ferdinand always seemed to know when they should stop to rest, and the fragrance of Taprobana or Molucca tea always prodded them into daybreak.

For most of the journey, the nebulous forms of the Middle and its encircling ocean sea were almost constantly in their sight, for no mound or bend of Yandu's branch eclipsed them for more than a few hundred paces. The Center appeared quite bright during the day, and the shrimp looked forward to its fine daylight as a welcome change from the eerie, subdued aurora of the great branch. Once within closer proximity, however, the Center itself was no longer visible because of the imposing moat, which towered far above them. But the glow of daylight from the Center easily penetrated the upper corner of the sea, so the day-night rhythm was never lost altogether.

Journey To The Middle

Eventually, they reached the end of Yandu's long branch, and, with it, they reached the ocean moat that encircled the Middle of the Primal Pivot. Up close it had a fresh, tingly silvery hue, rather than the distinctive golden aura it exuded from a great distance. Ferdinand and Guinness had, of course, both seen the moat before. It was, to be sure, the only moat they had ever seen. They assumed that it must be a quite typical moat, if in fact there were other moats elsewhere to compare it with.

But neither Ishmael nor Atollana had ever seen anything like it. The moat was a sea of dew-water without any wall to contain it, rising far above their heads and extending equally far beneath their feet, perpendicular to the branch. Although it was too vast to see more than a mere sliver of it up close, its flat, vertical perimeter appeared to entirely ring the Center with no vessel to hold it. The two shrimp thought it all very peculiar. Atollana submitted the only imaginable question :

"Why doesn't it all spill out?" she asked.

The two sheep dogs looked at each other and lightly tilted their heads. They looked baffled. Guinness wagged his tail only very slowly, while Ferdinand, careful not to be impolite, responded calmly to the shrimp question. His frustration was, however, evident :

"Why should it spill out? Do you wish it to spill out? It's a moat. Moats do not spill out. If it spilled out, it wouldn't be a moat. You've never objected to things not spilling out before."

"Oh, no, it's not that I object at all," assured Atollana. "In fact, I'm grateful that it does not spill out. It's just that, back where Ishmael and I are from, we have to bury our moats flush with the ground."

"Flush with the ground?" asked Ferdinand, startled. "You don't mean *literally* flush with the ground — ?"

"Well, yes, sir, indeed I do," replied Atollana, somewhat meekly. "We dig a trench and fill it with water. But only up to ground level, never above."

"Incredible!" snapped Ferdinand. "Absolutely incredible. Why would you bury your moats in a trench?"

"Well, otherwise," explained Atollana, "they'd gush out all over the place."

At this explanation there was an uncomfortable, motionless silence. Then Ferdinand and Guinness gave each other a droll, dumfounded looked which they always did when the shrimp uttered the incomprehensible.

"A bloody nuisance!" Ferdinand finally retorted, "having to bury one's moats. Is that really a veritable fact? Rude physics there on Earth. Don't you agree?"

Atollana was reluctant to assume that her observation about Earth moats was a *veritable* fact, as tea-time with Yandu had made her cautious about such suppositions. Perhaps it was a more pliable sort of fact.

As for Ferdinand, it should, he reasoned, have been self-evident that their moat was suspended in perpetual equilibrium between the two worlds it partitioned, being perfectly balanced between the powerful, opposing lures of the Primal Pivot and the extrum-Yanduian, *Mundus Mundanus* part of existence. In the mind of a sheep dog born to the Yanduian Highlands, it made utter sense for a moat to embrace such poetry. However, not anticipating any answer from the Earth-reared shrimp on the matter, he and Guinness pulled out the odd junk that looked like ocean debris and prepared the shrimp for their crossing through the moat and into the Middle. They took two rubber pieces that resembled webbed feet and stuck them on two of Ishmael's limbs. Each rubber flapper had a hole to put a foot in, although there didn't seem to be any way to choose which two of Ishmael's ten feet to stick them on (the matter was decided arbitrarily). Then they covered his face with a glass mask, carefully pointing it forward so that he could see out of it. Lastly, they took a device shaped like a long tube bent at one end, secured it vertically to the mask, and put the bottom, bent end into Ishmael's mouth. After hesitating for a

Journey To The Middle

moment, he tried to breath. To his surprise, the air found its way through the tube and into his mouth, although he couldn't understand why the sheep dogs thought it necessary that he use it.

"Well, the gear is all in order: fins, mask, and snorkel for your swim through the moat. We have only one set, so Atollana will have to hold onto you very tightly until you are through and into the Middle," Ferdinand explained.

Ishmael, it should be mentioned, was not comfortable wearing their swimming contraptions. He told the sheep dogs that, back in the Aquanesian Sea on Earth, shrimp swam quite well without any such apparatus to help them. But Ferdinand and Guinness made no attempt to respond to such a strange claim, and once more gave each other that queer glance reserved for outlandish shrimp statements.

"Now, have you any questions," asked Ferdinand, "before we conclude our herding?"

There were no questions. "If not," Ferdinand continued, "Guinness and I must hasten back to the forest. Once you enter the moat, we will be beyond contact. You will be on your own. Now, to penetrate the wall of the sea, gently press yourself into it. When you reach the far end, you will have to ascend the shore to get onto land, for as you can see, the Middle is higher than our present level. Much higher.

"One final, extremely important point," Ferdinand continued, "while you are within the moat, be sure to remember which way is up. Inside the moat, there is no way of knowing. You just have to remember. If you get turned around, you will come out on the underside of the Primal Pivot. We do not know what is there. It is unherded, for it is truly beyond the realm of all herding — it is the Antipodes."

Shivers quivered the shrimp's souls at the sound of the word 'antipodes'. But they had no questions and were ready to proceed.

"Then we must begin our separate paths now," continued Ferdinand. "You will be in our hearts always. It is unlikely that you will return by this way, if you ever have reason to leave from where you seek. It is doubtful that we will ever meet each other again."

Ishmael thought it was strange to say this, because there really didn't seem to be any other way to return to Earth and the Aquanesian Sea except by this same way they were travelling. While he was deciding whether or not to ask what they meant, Atollana climbed onto his back, holding securely as Ferdinand had instructed, and whispered that they should enter the moat without delay. "I cannot bear to say good-bye," she said, "so let us go now."

Ferdinand and Guinness carefully handed Atollana the parchment chart, which she clutched tightly, knowing how precious it might prove once they were unherded in the Center. The sheep dogs watched as Ishmael cautiously pressed his body against the dew and entered the towering moat. Dew-water quickly filled his snorkel, though this, of course, was of no consequence to Ishmael the shrimp, and he made no effort to swim upwards to allow the snorkel to pierce the top of the moat. The image of the two shrimp softly blurred in the water as the sheep dogs turned to begin their long trek back to the Yanduian Highlands. While Ishmael was learning to swim with the terra firmate gear in the strange moat, Atollana (whose vision was not constricted by a terra firmate diving mask) was distracted by a figure in the distance, appearing as if an apparition because of the soft distortion of the dew. So mesmerizing was the figure that Atollana decided to let go of Ishmael momentarily to be able to turn her body to see it better. But upon releasing her grip she quickly became disoriented, able to find neither the strange figure nor Ishmael again.

So Atollana continued swimming alone, apprehensive about the situation but reasonably confident that she and Ishmael would quickly find each other once they reached the shore of the Center. Ishmael, meanwhile, was trying to accustom himself to swimming in the dew sea, where his vision was gently blurred, and he had to negotiate his limbs differently than in regular water. He was entirely preoccupied with his formidable task of navigation and was not yet aware that Atollana was no longer on his back.

V
Passage To Terra Unherded

Much of the sheep dogs' concern and explanations were beginning to make sense to Ishmael. The moat had no bottom or floor, quite unlike his Aquanesian Sea on Earth. If he were to lose his orientation, there would be no way to know which side was up. There would be a fifty percent chance that he would step out onto the unknown, unherded, underside of the Primal Pivot. This was a novel problem which Ishmael did not enjoy. On Earth, no fishlet ever wondered which way was up, because the sea floor marked the 'down' side, while the waves and sun and moon and stars marked the 'up' side. And on Earth, gravity pulled you toward its center (a true-enough fact for this purpose), though the ocean floor was there to stop you long before you would reach it. But here in the ocean sea surrounding the Pivot, gravity let you differentiate 'up' and 'down' from all other directions, but not between the two. It pulled toward the precise center of the moat, equidistant between top and bottom, at which point there would be 'up' in both directions. Ishmael figured that the principle was probably the same as gravity on Earth, but that the aquagraphy was vastly different.

He found that there was, however, one reliable way to keep his course within the moat of dew. Because the moat was wider than it was deep, the dew depth (and thus the relative opaqueness) was greater side-to-side than it was top-to-bottom. Ishmael used this ninety degree dew parameter as a navigational device to keep his course. He hoped that perhaps someday he could tell the sheep dogs of his discovery.

After considerable swimming he found himself approaching a nearly perpendicular wall of solid rock, the sub-aquatic floor of the Middle. But as he swam upwards, it began to slope more gently and was cushioned by a sandy *terra*, in the same fashion as the sea floor rests on the shoulder of an island on Earth. He liked that.

Atollana had also reached solid ground, though she and Ishmael were not within sight of each other. She consulted the old parchment chart that the sheep dogs had given her, comparing it against the topography of the undersea wall of the Center, and was surprised to find it leading her in the opposite direction from what she had expected, a direction she would have dubbed 'down'. Believing she had gone astray, Atollana quickly turned back into the open moat and fastidiously followed the chart toward the Center, their common destination. That, she reasonably believed, was the surest way to meet up with Ishmael. Certainly a *platt* drawn by the great Pliny himself was more reliable than her own dead reckoning.

Meanwhile, Ishmael, reaching a fairly shallow part of the moat, now first realized that Atollana was no longer on his back. He nervously searched for her along the island's coastal shelf, but thought it unwise to re-enter the open moat until he checked for her in the Middle. Perhaps she had simply stepped ahead of him upon reaching shallow water.

Ondew

Anxious to find Atollana and to get his first glimpse of the Primal Pivot, Ishmael emerged from the moat, his diver's mask, snorkel, and flippers still securely fitted to his body. Carefully, he studied this new terra firma from side to side. But his mask was hopelessly fogged with dew. All he could see were gyroscoping patterns of color.

Now something extraordinary happened. As Ishmael stood on the shore trying to remove his muddled mask, a voice spoke to him. It was a soft and gentle voice. But though it was the purest and most disarming of voices, Ishmael was startled by this new presence. He also thought that what the voice said was very peculiar :

"You're naked! Oh, dear, a naked person!"

Ishmael was flabbergasted.

"Naked?" asked the shrimp.

"Naked," confirmed the voice.

"But why do you say I am naked?"

"Because you are naked, naked indeed."

As Ishmael tried to reason with his unseen accuser, his attempts to remove the mask became frantic. "But how can I be naked? I'm a *shrimp*! Shrimp can't be naked. Shrimp don't wear clothes."

The voice persisted. "But that's precisely why," it said. "You *are* wearing some clothes, but you haven't finished covering yourself. You see, I've heard about clothed people, and have been told about their proper customs. Now, Mr. Shrimp, if you don't want to be naked, you'd better remove those garment things from your body. In the meantime, I will turn the other way. I won't embarrass you. I've been told about embarrassment too, and know that people who wear clothes are embarrassed without enough of them. Tell me when you've undone yourself, and then I will face you again. I certainly don't wish to disrespect the customs of clothed people."

Ishmael fumbled, struggling to pry himself loose from the wretched swimming gear. Realizing that he would never get the mask off without all his limbs free, he first dislodged the flippers, then the snorkel, and, most difficult of all, he finally removed the confounded mask. Able to see again, he looked around, and promptly fell to the ground, confused and dizzy.

"Are you okay, Mr. Shrimp?" the voice asked.

Ishmael sat up and tried to gather his wits. He found himself in a world which for the most part was really not so unfamiliar. It was very much like a friendly tropical island on Earth, except

Passage To Terra Unherded

perhaps a bit more fantastic. However, toward the middle of the island, quite some distance, light seemed to dance in the sky, incessantly pushing toward what appeared to be the most inland point, groping for the island's precise center. Outside this interior region the sky was quite like an earth sky, though he sensed that light at certain angles still seemed to yearn for the center of the island. In the opposite direction, the world on the other side of the ocean sea moat was not visible, just as this world had not been visible from outside it.

Amid all his confusion, Ishmael had at first not even noticed that the creature who had spoken was there, facing away from him, only a few paces distant. She was a two-legged terra firmate, a human person. She seemed a bit anxious, and began to speak again.

"Please, are you okay, Mr. Shrimp? May I turn around now?" Ishmael was too baffled to reply. The terra firmate girl was still gazing off into the distance, but her body was beginning to twitter nervously.

"Well," the girl continued, "I am going to turn around now anyway, because I am worried about you!"

So the human person turned and faced Ishmael. She was slender, and somewhat short for a terra firmate. Her skin was of an olive color, and her hair, which was wavy and dark, hung in a disheveled fashion. She had a kind and peaceful aura. Dew was trickling from her swarthy body, giving a gentle sheen to her amiable, if somewhat peculiar, appearance. She was also, curiously, quite nude, save for the breadth of her hair, which dangled loosely past her waist.

Ishmael quickly mustered the same high degree of courage which Atollana had needed when she first addressed the sheep dogs. "My name is Ishmael," he began. "Ishmael shrimp. Are you going to hurt me?" he asked.

There was a brief pause before any reply came from the puzzled girl. She herself did not actually have a name (for names had not been painted until after Creation was painted), but remembered one which others had used for her.

"Perhaps I might also have a name someday," she finally answered (it was a racy thought for her). "But people call me Ondew.[13] Where is your friend — the one that was with you in the sea?"

Ishmael was startled. "You know about my friend?" he exclaimed.

"My dear Ishmael!" she replied, "Do you not see that I am dripping with dew? I was swimming about when the two of you first entered the water. But I got back here long before you draggled out of there."

(This was all very encouraging for Ishmael, who had wanted to ask Ondew about Atollana, but was wary. Ondew, however, seemed to be a very kind creature, which is not what he expected from a terra firmate. He reasoned that just as the girl's island was more of a *terra incognita* than a *terra firma*, Ondew herself must be a terra incognita firmate rather than the terra firmate he so feared. Perhaps, Ishmael reasoned, this was a significant difference.)

"I have lost her," he explained. "I am terribly worried about her."

Ondew smiled at the distressed shrimp and took his hand. "Come with me. We'll find her."

She led him to a path which wound around the rim of a jungle. While they were walking, Ondew tried to make her new friend feel more relaxed.

"By the way, Mr. Shrimp, I'm glad you managed to get that silly garb off. You looked so klutzy naked."

"I did?"

"You did."

---

[13] The name *Ondew* (sometimes *Undew*) apparently evolved because people referred to her simply as she who would *undo* Nothingness, the latter syllable being poetically transcribed as 'dew'. It is not known whether the mythical *undine* was inspired by it. — ED.

"But, Ms. Ondew," Ishmael objected, "you are not wearing any clothing. Are you not naked yourself?"

"Naked? Ishmael! I am no more naked than you are, now that you've gotten those things off your body. Only clothed people can be naked, yes? Though I don't know much about these customs myself, I've been told that certain creatures wear clothes, and are considered naked without them, while other creatures do *not* wear clothes, and are therefore not thought of as being naked."

Ondew was quiet for a moment as she tried to remember the cultural specifics of this curious phenomenon called clothing. The painted world, the girl reflected, had indeed concocted some intriguing customs and rituals. "From what I understand," she finally explained to the confused shrimp, "clothing is used to endow one's body with a *special sensuality*. You see, an unclothed body is just a neutral form. But to garnish one's body with clothing is to pepper oneself with the spice of eroticism. That is why people who wear clothing are embarrassed without enough of them — it is because they are ashamed to be viewed in a non-erogenous state."

Ishmael at first suspected that the girl had her ethnological facts entirely reversed, but he carefully and logically thought through the clothing-custom and realized that her treatise on the subject made unequivocal sense. It was obviously he who had remembered wrong. The girl continued her observations :

"Now do you understand why I was concerned about looking at you? You were being a clothing-type person, but without having fully covered your body. That is to say, you were naked." As Ondew contemplated all this, daring, impetuous thoughts danced through her mind: perhaps someday, just for a beautiful, innocent delight, she would try enshrining her body with clothing. But she quickly left those tasty thoughts to a more private moment. "I guess I myself am too shy," she finally confessed to Ishmael, "to cover my body with clothing."

"Me too," admitted Ishmael. "I get embarrassed even without them."

"Well, if you get embarrassed as it is, it's a very good thing you don't wear clothes, or it would be even worse," Ondew speculated. "Do you agree? I'm told that these are Cultural Truths, which are a kind of fact. And facts are, of course, *facts*." The girl was then silent as they wound around various botani/ geological features of the jungle. Ishmael, studying her expression, noted that her facial muscles were slightly tense, as though her last statement had not yet been relieved of duty.

"Funny, though," she finally continued, "I never painted any facts. I wonder how they got painted?"

Ishmael was too preoccupied with the question of Atollana's whereabouts to ponder what the girl was talking about, or even to wonder whether Yandu knew about the 'Cultural Truth' genre of fact. More important matters took precedence; the jungle path had led them to another point along the moat shore, but there was no sign of Atollana. Ondew tried again to comfort Ishmael, and suggested that they now head up a certain hill to survey the terrain. If that fails, she said, they would re-enter the moat and venture to the underside of the Middle. That, of course, was the unherded Antipodes of the Primal Pivot. The possibility that Atollana may have gotten turned around and ended up there horrified the shrimp.

"What is her name, Ishmael?" Ondew asked.

"Atollana. We've travelled together from quite far away."

"You mean from another part of Everything? From *Mundus Extrum Yandum*?"

"From the Aquanesian Sea. On Earth. We live in the Aquanesian Sea on Earth."

"Oh," answered Ondew, trying to make sense out of his answer. "What kind of plant is Earth?"

"Well, a big round one, of course," Ishmael explained.

"Round?" asked Ondew. "You mean actually spherical? What kind would that be? A guava?"

Ondew focused her eyes toward the sky for thinking purposes. "Maybe a persimmon? Or more likely a pomegranate, I suppose . . ."

Ishmael, realizing that Ondew lived on a garden island, clarified that Earth was not really like a plant, but rather that it had a hard exterior shell, not like that of a shrimp, but more like that of a coconut, although, to be sure, it was not a coconut. Ondew then narrowed the possible choices of which plants Earth could be.

"You mean it's a lychee?" she deduced. "Or a pineapple? No, pineapples aren't round. Could be a jack-fruit or a bread-fruit, of course . . . ".

Finally, Ishmael had to explain that Earth wasn't a plant at all, but rather that it was basically just a big old rock. Ondew was stupefied to learn that there were people, her new friend among them, who inhabited such a desolate sphere. "You live on a boulder?" she asked, staring into Ishmael's eyes. "Rocks are beautiful, of course, but I didn't realize you could make one your home."

She gazed far away for a moment and whispered something about never having painted such an idea. Then she looked back at Ishmael and tried to envision what this dead Earth plant would be like.

Her words were shaky. "Living on a stone," she asked, "are there grey clouds and pink skies, do trees blossom, is there wonder, can people love each other, is there music and poetry, and can one dance?"

Ishmael assured her that on Earth there were all these things, and that Earth was in truth not a lifeless boulder at all. It was, he elaborated, a world of unfathomable beauty, and one which was brimming with all sorts of wonderful life. He first tried to simply enumerate all the creatures which inhabit Earth, but quickly realized that Ondew was still imagining the planet itself as a mere chunk of granite. The girl's concern inspired him to explore new terms with which to explain Earth. Earth, he told her, simply used rocks of various sorts as a kind of backbone, perhaps not so unlike her own terra firmate spine and ribs. Thinking back to some of La Vecchia's more impassioned talks, Ishmael tried to explain that this rock foundation unified bountiful forests and vivacious jungles, eternally frozen tundras and vast deserts, immense oceans and tiny ponds, flowing hills and imposing mountains. He told her that Earth was more diverse than he could possibly articulate, yet it was all unified by the rock inside.

Nor was this spine itself merely stagnant rock. Earth, in fact, had quite a complicated fruity inside, including pulpy parts and viscous sections, and even brilliant red juicy stuff which would sometimes spurt out when over-ripe parts got squeezed. It was in this manner, Ishmael added, that an isle such as hers would be born in the oceans of Earth.

Ondew was particularly curious about this island-birthing occurrence on the Earth plant, and questioned him in detail about it. From what she could deduce, it began with torrid, aroused rock deep in Earth. So agog would this rock be, throbbing inside the beautiful Earth, that it would ultimately explode, spurting its virile, molten seeds into the sea-womb of Earth, thus conceiving new, fertile islands. Ondew's eyes were unshakably anchored to Ishmael's, so taken was she with this beautiful description of how Earth perpetually and spontaneously sculpted itself anew, a world in the midst of unceasing experimentation, unending rebirth in endless variation. Creation, all by itself, had invented regeneration.

Ishmael continued by describing Earth as a collective entity on which myriad creatures lived, all of whom shared one trait: each was dependent upon both the planet and the rest of her creatures. Every being, however obscure (though 'obscurity' was a concept Ondew could not understand despite Ishmael's repeated explanations), was a vital link in a system as complex as it was fragile, and it was the sum of all these which was called 'Earth'. He was happy to add that this made Earth, in a sense, a life unto itself and really not so dissimilar to her jungle island.

Ishmael had never thought about Earth from this perspective before, and he liked doing so. Ondew thought it all quite wondrous, and no longer worried about people living on a rock; she was, in fact, radiantly happy that the painted world had been so imaginative and inspired in its pursuit of beauty.

But what of this strange word 'planet' that Ishmael used? Ondew had, at first, assumed that the tired shrimp was mispronouncing 'plant', but his continued and careful enunciation of it as two syllables made her suspect that it was a new word. Now that her rapacious curiosity about Earth had been tolerably satiated for the moment, she asked him about 'planet' and discovered that it was indeed a different word from 'plant'. A planet, she learned, was a life orb which had developed into such a contrapuntal megafugue that the *sum* of its voices had become a life in itself. Thus 'planet' is to 'plant' what 'polyphony' is to 'voice'. It was all really perfectly natural and utterly beautiful.

How did this interesting word evolve? Ondew surmised that the sound '*e*' — the first tone in '*Earth*' — had been placed inside the word *plant* to differentiate an unaccompanied being — such as an individual fern — from a massive, contrapuntal one, of which the Earth plant was a singularly masterful muse. Ishmael knew nothing of the word's etymology and couldn't corroborate her theory.

Finally, Ondew made the observation that her own island, like Earth, was indeed a contrapuntal plant, and her *terra*, like the Earth *terra*, was coddled by an ocean sea. Enamored by the parallel, she smiled at Ishmael, dug her hands into loose ground below them, and spoke of the dirt in her palms as 'earth'.

The girl had innumerable questions for the shrimp and wanted to hear much more about this Earth, but she knew he was distraught about his friend Atollana. Their path had now brought them to the observatory hill she had mentioned (it rose at such a sharp angle that Ishmael secretly dubbed it a mountain). They followed a trail up toward its pinnacle, which was flanked by a tiny valley of sorts. Upon reaching a ridge near the summit, Ondew sat down next to a small fig tree which straddled the bank of the little valley, and instructed Ishmael, who was looking aimless, to sit beside her. Once the shrimp had nervously taken his place, the little tree offered each of them its fruit.

From their mountain vantage point, the two were now able to inspect a large part of the ocean sea. Ishmael confirmed what had seemed evident from a great distance back on Yandu: this moat, though entirely surrounding the Primal Pivot, was not touched by anything on the outside, and was accessible only via Yandu's branch. This further worried the poor shrimp, who considered the chance that Atollana might attempt to return to Yandu rather than the Middle; for if she were to do so, and if she were to miss the narrow branch of Yandu, she would emerge from the moat into deep space instead.

Ondew was much more relaxed, and assured him that there was no reason to worry. Peering out over the mountain as she ate her fig, she scanned the topography for Atollana, and spoke to Ishmael without turning her head. "Either she swam far around the moat before climbing out, or she got switched around and went to the bottom side. If she peered out into space from the edge of the moat, she wouldn't fall. And the trees will care for her if she has simply gone beyond our sight. But probably, I think, she is in Zinnia."

Ishmael didn't ask where Zinnia is. Instead, he wondered which trees would care for Atollana if she were lost in the jungle, whether it would be the more earthy and earthly looking trees, like the banyan or mangrove or pandanus, or the towering trees far in the island's mountainous interior. Ondew, in the meantime, got up off the ground, smiled at the fig tree, and took Ishmael's hand. She led him to a precipice of the mountain which extended far out over the moat, far enough that they overlooked the open, floorless part of the sea, rather than the underdew shoulder of the island.

"Hold onto my hand very tightly, dear Ishmael," she whispered, "and jump off with me when I count 'seven'. From this cliff we are at just the right height to give us enough momentum to get past the center point of the moat when we fall. This way, we can continue the same direction and rise up on the other side without our slightest effort. And shall we agree to dive head first? If we dive head first we will be the right way up. You see, their feet and our feet join against each other."

Skipping the six preparatory numbers that Ishmael had expected, Ondew yelled 'seven' and pulled them off the ledge (Ondew had selected the number 'seven' because its two syllables allowed for both a push and a release of the feet). Ishmael saw her gather a deep breath as they fell, and remembered that terra firmates cannot breathe in water. Just as her lips sealed, closing her chest and cheeks tight with air, they hit the soft tingling dew-water. Piercing the moat surface head first as Ondew had instructed, they plummeted deeper and deeper into the ocean sea toward the Antipodes. Their momentum slowly decreased until, just as their downward motion was about to be overpowered by their upward buoyancy, they passed the mid-point of the moat; which is to say, they passed the precise level at which the buoyancy of their bodies shifted 180 degrees from above to below, or rather the point at which 'up' and 'down' got switched around and their descent became an ascent to the underside. 'Up' (formerly 'down') being the desired direction, they allowed their bodies to float lazily to the antipodean surface.

Thus one could pass from the 'top' world to Zinnia (and presumably visa-versa) with a single, effortless gesture. By falling freely into the moat, the body became charged with just enough downward energy to counteract one's natural bouyancy as far as the mid-point of the moat. At the exact middle, one's buoyancy reversed, the body then continuing in the same direction at a gently increasing, rather than decreasing, speed. It was, Ondew thought, rather like a modulation in music,[14] in which an altered, leading note (full of energy) would ferry the

---

[14] In music, the establishing of a new key or pitch center. — ED.

music to a new key, a new world, where that very same nervous, altered note, would then float by itself.

The two glided parallel to the cliff which formed the underdew ledge of the Center, following it straight up through the water to the Antipodes. Ishmael had expected that it would begin to slope toward the underside as a sandy shore, but instead it continued as a sheer rock. When they reached the antipodean surface of the moat, he saw that it was a precipitous mountain which rose far into the sky.

"Follow me, Ishmael," instructed Ondew, "we must find the path." Treading dew, they scrutinized the mammoth cliff towering before them. Ondew motioned toward a narrow crevasse which, upon approaching closer, revealed a long, slender staircase carved into the rock. They pulled themselves out of the moat and began climbing the stone stairs up the mountain. Reaching the top, they were on an immense, barren plateau rim, beyond which the land again climbed steeply.

"Outer Zinnia," announced Ondew.

"Zinnia?" asked Ishmael.

"Zinnia," confirmed Ondew. "That's where we are. Outer Zinnia. When we reach the top plateau of the island, we will head inland and eventually cross over a deep gorge. The gorge is called Aqua-Abyss. It encircles the Kingdom."

"Kingdom?" asked Ishmael.

"Kingdom," assured Ondew. "The Kingdom of Prester Prawn. Of Aqua-Abyss. In Zinnia. Inner Zinnia. 'Abyssinia' for short."[15]

The two happened to glance at each other as they continued climbing. Ondew shrugged her shoulders.

---

[15] As noted previously, The Kingdom of Prester Prawn is perhaps a corruption of The Kingdom of Prester John, a mythical Christian kingdom, long believed to lie in Africa in Abyssinia (hence Aqua-Abyss in a land known as Zinnia). Originally, however, the kingdom was believed to be somewhere in Asia, and its history is closely entwined with the Crusades. — ED.

"The Kingdom of Prester Prawn of Aqua-Abyss in Zinnia," she summarized. "I didn't paint it. I never painted any kingdom. It must have gotten painted by someone who got painted by someone who had gotten painted. Or maybe more than that. I really don't know."

Ishmael didn't know much about painting, and didn't ask how they were to cross over a gorge. He just kept climbing, trying to keep up with the girl.

VI
Zinnia

They persevered for quite some time, and eventually reached the gorge that Ondew had described. It was about 200 terra firmate paces wide, Ishmael figured, and at least as deep. But the Abyss proved to be quite unlike the barren, empty crevasse that Ishmael had envisioned, for from its bosom erupted rivulets of a golden dew-mist which danced wildly throughout the canyon.[16] Silky, silvery clouds of the dew would alternately settle upon the Abyss floor, then shoot up flippantly while other dew embodiments, in turn, gently glided to its hazy depths to rest. The graceful whimsies of these formless dew nebulae roamed the expanses of the Abyss; and as this vast gorge, like the ocean sea, completely encircled the Kingdom, the glowing, misty aurora appeared as an extraordinary halo extending fully to the horizon.

Ishmael looked at Ondew in hope of an interpretation of the wonder. The terra firmate girl spoke without moving her eyes from the canyon.

"You see, Ishmael, in order to paint one, the other got painted," she explained. But now the shrimp looked even more confused.

"It is the Great Contradiction," she tried again. "What you see is, of course, a meteorological phenomenon. But in Zinnia, the Aqua-Abyss is a metaphor. It is poetically considered the place where life comes to terms with suffering. It is here that they meet. Suffering did not *want* to get painted. But you can't

---

[16] The "dew-mist" in this passage was probably steam or vapor of some sort, perhaps moisture-laden fog spun about by winds within the gorge. — ED.

paint just one. If you paint life, suffering gets painted also. That's what happened. Now, come, Ishmael, let us begin to sing."

"Sing?" asked Ishmael, "But I don't sing. I cannot sing. Why must I sing?"

"Well, to cross the Abyss, of course," she said. "We need enormous energy to reach Inner Zinnia! Music is energy. That's what energy is. That is what life is."

Ishmael tried to be accommodating and offered to attempt to sing something, but Ondew was not interested in this. "You cannot sing as a chore," she retorted, "nor can you just sing 'something'. Now, come here and hold tightly onto my back. I will sing for the both of us, for I do love you, Ishmael Shrimp, and together we will sing our way over the Abyss. Come, put your arms around my body and grasp firmly!"

The shrimp felt somewhat embarrassed to subject Ondew's soft human skin to his inelegant (if practical) crustacean claws,[17] but she gently grasped one and placed it against her body to make him feel more confident. As he cautiously put some legs around her waist and shoulders, she turned her head and smiled at him; this comforted him greatly. She then began to sing a song, a divine song. Ishmael's limbs, particularly those which grasped her chest, began to vibrate softly with her singing, and in this manner he felt her voice through his body. He became consumed by her song, and, without realizing it, began to sing with her. This, of course, helped their upward force. The golden dew mist began to swirl about, the ground began to release its hold over them, and they slowly glided over the staggering Abyss. From midway over it, Ishmael peered down and saw that these dew forms assumed no particular shape, and were only passingly autonomous, oozing in and out of each other. As they approached the chasm's inner shoulder, the banks of Zinnia proper, Ishmael thought about what Ondew had said. Perhaps life and energy could not exist without music. Perhaps they were one and the same. And perhaps it all had

[17] An error in the text — shrimp have swimmerets, not claws. — ED.

something to do with the decay of the Aquanesian Sea and their reef.

Soon they came to rest on the far side of the Abyss, which is the inside of the Middle, the kingdom known as Zinnia. Ondew finished singing, and then Ishmael, finally realizing that he was also singing, stopped. They both surveyed the land.

"Zinnia," stated Ondew, categorically.

"*Zinnia*," stated Ishmael, categorically.

And so they were indeed in Inner Zinnia, quite different from the austere plateau they had just left. The terrain of Zinnia was gently rolling, with scattered brush here and there. In the distance, the vegetation seemed to become thicker and more varied. A bit beyond that, Ishmael could make out groupings of what appeared to be many little thatched huts, not so dissimilar to descriptions he had heard of the dwellings of jungle and forest peoples back on Earth. What was extraordinary, however, was that scores of figures, of varying sizes but generally small, appeared to be floating about at all different heights. Ishmael squinted sharply in hopes of discerning them better. Ondew, meanwhile, identified the figures for him.

"Zinnians," she explained.

"*Zinnians*," Ishmael accepted.

The citizens of Zinnia, apparently, did not get about by touching the ground, nor even by flying. Rather, they flittered about, swimming as if they had a sea in which to swim (which they did not). In this respect, and in the array of marvelous plants growing on its surface, Zinnia resembled the reefs of the Aquanesian Sea back on Earth.

Ishmael, obviously curious, asked Ondew how these citizens of Zinnia stayed buoyant.

"By energy," she replied flatly, as if her answer were entirely self-explanatory. But glancing at Ishmael, she noticed that he still looked unanswered, and so offered a clarification.

Zinnia

"That is to say, Ishmael, by *music*."

Listening carefully, Ishmael indeed heard the soft murmur of marvelous music. Ondew explained that in Zinnia there was eternal song, song always to be heard, for in Zinnia no creature knew the breath of life without music. They hovered to song, swimming via the buoyancy of their chant. In Zinnia, music and energy were still one, and together they were called Life.

In contrast, Ishmael and Ondew, already peculiar looking creatures in Zinnia, were particularly conspicuous because their feet touched the ground. Ishmael was self-conscious about proceeding any further toward the nearest Zinnian villagers in their unlevitated state. He asked if perhaps they should sing.

"Don't worry," Ondew assured him, "There is no need for us to imitate them. Many here like to walk on the *terra* as well, if that is their music. In Zinnia, no one asks why you are as you are. No one judges your chant. And anyway, they've seen foreigners before."

Presently a squid-looking fellow approached. He wafted toward them, about shoulder level to Ondew, which was above Ishmael's head. His tentacles softly pulsated to the chant of his quiet and wordless song. The energy of the squid's song turned into a smile on Ondew's face, and then his song bore words.

The words were :
>*I sing up a fourth,*
>*up a minor third,*
>*there to trill a bit,*
>*and add a word.*
>>*Like mushroom, bamboo,*
>>*perhaps Katmandu.*
>*Jump down so deep,*
>*first an octave leap,*
>>*then a minor second too.*

Then the squid continued singing softly, but again without words, and Ondew answered him in song, harmonizing the squid's non-verbal chant.

Ondew sang to the squid :

We search for Atollana,
a guest of the dew,
who perchance has wandered,
wandered through,
    she came from the unknown,
    Yandu and beyond.
    Perhaps you have seen her,
    a shrimply vagabond?

The squid replied :

A chart of old
A chart with secrets told
and mysteries to unfold
With bold letters in gold.
    Though from the ocean still cold,
    and half rotted from mold,
    this chart she unrolled
    in quest of the alluring place it told.
But dear friends be consoled
for communal song I now will hold,
to warn all in our chantly fold,
that misfortune need be controlled !

The squid bowed politely and zipped away. "She's been by here," Ondew translated to Ishmael, "and is apparently heading for the Center, following a chart or map of some sort. This kind Sir Squid is going to inform all the villagers of the situation, and with their help we will surely find her."

So the squid person, whose name was Mr. Mook,[18] fetched his son, a handsome and tawny youth named Calamito,[19] to lead

---

[18] Mr. Mook's surname apparently (?) derives from the Thai word for 'squid'. Assuming this etymology to be correct, the word rhymes approximately with 'took'. — ED.

them to the Center. They began their journey without delay, Ondew and Ishmael walking gracelessly on the ground, following their young squid guide. The youngster explained that they would reach the Chanting Jungle by nightfall.

The Chanting Jungle

The Chanting Jungle lay unevenly scattered about the perimeter of the Center. It was a jungle of lush and wonderful growth, some of which anchored itself in the ground, some of which hovered above unattached, softly chanting if it wished, slowly dancing if it wished. It was, further, a very friendly and hospitable jungle.

By dusk they reached its outskirts. A plant with a reddish-green stem and several white-ish (and very elegant) flowers chanted over to Ondew, shyly wrapping itself around her arm and gently tugging as a signal to follow it. Calamito explained that this was a Dorian plant, and that it would escort them within the Chanting Jungle.[20]

They followed after the Dorian plant, working their way into the thicket of the Jungle and reaching an open area of very low-lying green plants bearing countless tiny golden-yellow flowers. The squid identified this as an Aeolian grove.[21] Here

---

[19] Calamito must have been a Zinnian colloquial term for a young male calamare (as squid are known in Iberia and the Mediterranean region). — ED.

[20] 'Dorian' is a musical mode, or scale, one of several whose history extends back through medieval Europe to classical Greece. Plato, Aristotle, and Ptolemy of Alexandria all refer to the Dorian mode, as well as the Phrygian, Lydian, Mixolydian, and others. Dorian is the mode whose tonic would be 'd' with natural pitches. Vestigial influence of the Dorian mode can be found in Bach, whose first (g minor) sonata for unaccompanied violin bears only one flat in the key signature, technically placing it in the Dorian mode (Bach inserts the "missing" e-flat as an accidental). A well known folk song constructed in the Dorian mode is "Greensleeves" (though some modern renderings lower the sixth of the scale to transform it into the standard 'minor' scale). — ED.

[21] In music, Aeolian is the mode whose tonic would be 'a' with natural pitches (i.e., what is commonly known as a natural minor scale, although that scale's history is also entwined with the Dorian mode). — ED.

the Dorian plant pointed to a corner of this grove where a large Aeolian fern was waiting to greet them.

The fern explained that an unlevitated shrimp damsel in search of a friend, guiding her way with what appeared to be a geographical *tabula*, had reached the Chanting Jungle earlier. Although the shrimpess was anxious to continue on to the Center, for she believed that her companion was already there, the Elders of the grove — all of them utterly sensible Gregorian ferns — submitted that they probably would have known if he had passed through. They implored her to take refuge in the Gregorian Grove for the evening, to which she had reluctantly consented.[22]

So Ishmael's hopes were piqued as the Aeolian fern chanted a message toward the Elders of the Gregorian Grove with the news that the second shrimp had safely arrived at the Jungle. Soon a fern appeared through the Aeolian thicket leading a very tired-looking shrimp lass. Atollana (for yes, of course, it was she) rushed over to Ishmael, the two embracing with reciprocated words of relief as squid and plants chanted songs of happiness.

Atollana exchanged smiles with Ondew, and realized that this terra firmate woman was probably the figure who had caught her attention in the moat. So drawn was she to the girl that the shrimpess even forgot her deeply rooted cultural fear of terra firmates. Ondew wondered for the first time what reason these two harmonious people had for travelling so far from their home.

The Dorian plant told them that they were to stay in the Aeolian Grove for the night. "I will ask a Mixolydian fern bush to check on you and offer berries for your dinner," said the plant, adding that they need only chant should they require any help.[23] "The Aeolians will pass your chant in the right direction

---

[22] Clearly, this 'Gregorian Grove' is somehow tied to Gregorian Chant, the medieval church genre. — ED.

[23] In music, Mixolydian is the mode whose tonic would be 'g' with natural pitches. — ED.

Zinnia

to reach whichever grove would best assist you," the Dorian explained as it curtsied and quietly chanted away.

So the four of them — Ishmael, Atollana, Ondew, and Calamito — stretched out for a much-needed rest. The shrimp, in particular, were exhausted. There were many adventures which the two were eager to share, but only one question was so nagging that it could not wait for the morning: Atollana was perplexed by the fact that she had reached the Jungle before Ishmael. She had continued on her own because she assumed that Ishmael *must* have been ahead of her. Though Ondew and Ishmael tried to reassure her, the answer to her question was horrifying. By having followed Pliny Plankton's chart so meticulously, she had taken them all to the mysterious, unknown underside of the Center, the very place which Ferdinand and Guinness had warned was beyond the realm of all herding. They were in the Antipodes.

Atollana's earlier relief now clouded with concern, Ondew comforted her. She assured the troubled shrimp that although Zinnia was perhaps unknown (she did not use the term 'unherded') beyond the Center and her own little isle, it was certainly known to *her*. And Ishmael reminded her that it was also, evidently, the place Pliny Plankton had reached. Nonetheless, the Aeolian ferns, sensing Atollana's troubled mind, helped gently prod the shrimp to sleep with their mildly inebriating fragrance.

Ondew and Calamito, in contrast, chatted about the Chanting Jungle. The two sat up much of the night listening to the wonderful counterpoint to be heard in the groves, and discussing the myriad sorts of marvelous polyphonies that were possible in such a colorful and diverse jungle. Ondew, for example, speculated about the resplendent counter-melodies that might be had if a Mixolydian fern and a Dorian plant were to chant together above the pedal drone of the Aeolian grove. Calamito's eyes glistened at the inventive idea.

"But rarely done, of course," cautioned the squid chap. "Much too immodest. They would only do such a thing one day a year, on the Fugafest, the day when Zinnia is aligned with the axis of the Universe."[24] On that day, he explained, all proprieties of fugal etiquette are abandoned, all customary inhibitions of counterpoint are ignored. On that one day the traditional taboos of melodic relationship are laid aside, and the plants of the Chanting Jungle, and indeed all the creatures of Zinnia, may explore their deepest fugal fantasies. Ondew was a willing audience for piquant tales of this phenomenon called Fugafest.

There was, to be sure, a veritable cornucopia of rumors and gossip relating the contrapuntal mischief associated with the musical feast, and Calamito was quite happy to share some of the spicier Fugafest legends with Ondew. It was said, for example, that one year two succulents successfully maneuvered a polytonal retrograde of their original motif, followed by a lengthy episode filled with delicious Levant-sounding augmented seconds (Ondew was not privy to the terminology but easily understood the concepts once explained).[25] Another year some reeds performed a six-part fugue in which alternate statements of the motif were played backwards and inverted, and some episodes occurred simultaneously with the fugal statements. At yet another Fugafest, an entire valley was awash in fugues built exclusively on quarter-tone motifs.[26] It will be no surprise that although some (notably the jelly-fish) seemed to adapt well to the quarter-tones, others (especially the cacti)

---

[24] Fugafest (**foo**•ga•fest) was apparently a festival based on the fugue, a complex form of musical counterpoint. There are two basic characteristics of the fugue which are relevant to the *Codex:* all voices are of equal importance, none being subservient to others, and all are built upon the same basic motif. In music it is generally noted in the Italian spelling *fuga*, hence *fuga + fest* (festival) = the Zinnian *fugafest*. Symbolically, coming on the day when Zinnia was believed to be "aligned with the axis of the Universe," Fugafest apparently celebrated the Universe's being at once infinitely diverse and ultimately one. — ED.

[25] In music, an augmented second is an interval equivalent to three half-steps, e.g., f to g$\sharp$, and is a common element in music of the Middle East. — ED.

[26] A quarter-tone is the interval of one quarter of a whole step, or one half of a half-step (the half-step is the smallest interval normally used in Western music). — ED.

had a terrible ordeal with them, though all were exhilarated by the effort. There was an opulence of such stories in Zinnian lore.

"Really quite a licentious affair," Calamito reflected. "At the beginning of Fugafest, most are careful not to attempt any counterpoint which they cannot confidently execute, but by the end some will brave fugal maneuvers which are so perilous that they frequently end in utter disaster. All in great good cheer, of course."

Fugafest, Calamito continued, coincided with the one day of the year that the *terra* trembled. On this day a fabulous foggy mist permeated Zinnia, adding a dash of impressionism to the affair. So absorbed did Zinnia become in the Fugafest that over the eons a custom had evolved through which Fugafest was ended and the gyroscopic stability of the Center returned. "At the close of the day," the squid explained, "everyone happily exhausted, the old plants of the Gregorian Grove restore tranquility by a very low and timeless chant." After many unabashedly risque stories of Fugafest, Ondew and Calamito fell asleep.

When the dew brought morning to the Chanting Jungle, a nearby bamboo orchard gently woke the crew with a flutey melody neatly balanced above the reedy drone of elder bamboo. And when the four sojourners sat up and opened their eyes, the scent of Mixolydian tea awaited them, prettily prepared in elegant bamboo mugs.

(A certain Zinnianism needs to be explained here. When Zinnians sleep, they do not cease chanting. But their asleep-chant emanates from their dreams rather than their conscious thoughts, and so Zinnians in slumber — excepting for the occasional odd nervous dream — tend to hover quite low to the ground. However, squid guide Calamito, having dreamt intensely to digest the day's considerable events, inadvertently wafted away a bit and awoke in a neighboring grove. The others were still in sight, however, and he quickly joined them for tea.)

The Crustacean Codex

During tea-time Ondew and various Zinnians queried the two shrimp about a matter that was too obvious to have occurred to them: now that they had found each other, would they still like to continue to the Center on the Zinnian side? Or perhaps return to the other side and approach it as they had originally intended? Ondew warmly welcomed them to visit her side, adding however that she herself had not been to the Zinnian Center for a very long time. They were now so close that she would be happy to join them and guide them back to her side afterwards. The shrimp had sufficiently lost their terror of the word 'antipodes' to consider this kind offer with fresh curiosity (there was also, they admitted, something bewitching about being beyond the reach of any and all herding). They solicited the advice of their hosts, who in turn, hoping to advise them more wisely, asked what the purpose of their journey was. But that unassuming question caught Atollana and Ishmael vocabulariless. They tried to explain that their arrival in Zinnia was the chance result of Atollana's painstaking adherence to Pliny's chart following their separation in the moat, which of course followed Yandu's tea-time conversations and their inexplicable joining with the dew, and that it all began with the terra firmate girl's enigmatic verse etched on the shell. But as to the purpose of their travels? Atollana recalled the desire to see their world from the distance of the birds. But after hearing herself say this, she added that perhaps, in the end, it might also be to see it from the intimacy of the dew.

After much discussion and much Mixolydian tea, the shrimp made their decision. They thanked the various plants of the groves for their hospitality, and set off for the Zinnian Center. All said it was a cherished place. And, further, they learned that Pliny Plankton was usually meandering about somewhere near there.

During their hike, the two shrimp had ample time to chat about what had happened to each of them since their separation in the moat. Ishmael was particularly curious as to how Atollana had managed to cross the Aqua-Abyss without the advantage of a native guide. She was, needless to say, thoroughly mystified when she reached it. But, scrutinizing Pliny's chart for an alternate route, she noticed two curious things.

Directly over the gorge, the map was embellished with a little vignette depicting what appeared to be a battle. As best as Atollana could interpret it, two figures, one labelled *'Beauti and Life'*, the other *'Suffering and Deth'*, were clashing, each one's hands wrestling the other's. Both figures seemed androgynous, and both had pronounced, nearly stylized, expressions. The first appeared strong, gentle, and surprising peaceful, while the latter appeared desperate, doomed, and even repentant, as though this figure, though the aggressor, had somehow never wanted this confrontation nor even, perhaps, wanted to exist.

The other detail she found on the chart was an inscription, partially rubbed, but nonetheless still legible. It advised travellers who had reached this place to make music *'by your very voice or other means true to your chant'*.

"I felt so dejected when I reached the canyon," she commented, "and the chart's little picture was so disturbing. The dew seemed so intimate, and I so lonely and frightened, I was quite pleased to take the chart's instructions. I started singing, really to console myself, but then found myself lifting off the ground and crossing to the other side. A better chart I could scarcely have imagined!"

VII
Zero Degrees On All 'Tudes

By midday they reached a clearing in the jungle. In the middle of the clearing there was a ring of very ancient bo trees, whose trunks formed a low circular wall of sorts.[27] The limbs of the trees were sparse and generally branched outward from the circle. The space within the circle appeared to be hollow. But what was strange was that nothing, neither the trunk walls nor the ground itself, was visible within the circle below ground level.

Very near it were two old, bearded, thoroughly disheveled men, both terra firmates and obviously not native to the Kingdom. They were sitting (unchanted and unlevitated) on a rock and facing opposite directions from each other. The closer of the two wore a tattered golden orange robe and held a ragged old driftwood cane; the other wore a frayed grey garment and supported a large book on his lap. Both men greeted the arriving entourage of shrimp, terra firmate, and squid.

"Ahh, welcome, welcome, my fellow prisoners," the closer one announced, as the other nodded in greeting. The pilgrims were flustered by the strange salutation, but none questioned it at first. Instead Calamito, being the only indigenous Zinnian in the group, introduced Ondew and the two shrimp, and explained that Atollana and Ishmael had travelled to Zinnia from the faraway world called Earth.

---

[27] This and subsequent references to the bo tree are probably meant figuratively; the bo was the tree under which Siddhartha Gautama (Buddha) attained enlightenment. Although transliterated as 'bo', the consonant is closer to the English 'p'. Also spelled as bodhi. — ED.

Zero Degrees On All 'Tudes

"And I am Prester Prawn," the orange-clad of the two old men rejoindered.

"He is Prester Prawn," the other man continued, closing his large book and standing up, "King of the Kingdom of Aqua-Abyss in Zinnia. I am Pliny Plankton, an historian, and such is as it is stated in the Histories. Though human in body yet I defer to the simple dignity of the plankton."

Ishmael and Atollana were, of course, in awe at having found the great Pliny Plankton the Elder. However they were equally confounded by being referred to as 'prisoners'. Instead of revealing at once that they had been raised by his old friend and scribe, La Vecchia (whatever her real name was), that they held the necklace he had left with her (Atollana concealed it for the moment), or to break the news of the anemone's death, they stood speechless. Pliny Plankton, caught by the shrimply silence, realized that perhaps his preamble had been too curt.

"My colleague," Pliny elaborated to their Extrum Zinnian guests, "is Prester Prawn, King of the Kingdom of Prester Prawn of Aqua-Abyss in Zinnia. That is, 'Abyssinia', Prester Prawn thereof. Though in human incarnation, yet in modesty he remains a prawn. He wears his orange robe and pokes about the bo trees in hopes of a more elevated understanding."

The Prester himself to tried to clarify further :

"Yes, but mine is a pathetic story. I wanted to be the poorest and most humble in the Kingdom, so thus I would truly be King. But when Sir Pliny arrived, he duly informed me that there were neither rich nor poor, neither famed nor obscure in Zinnia, and thus I was deprived of being the poorest and most humble."

" 'Tis his tragedy," interjected the squid.

" 'Tis a penance, say some," added Pliny Plankton, "if one considers his Storie. And I have chronicled this Historie for future interpretation. It happened thusly :

The Crustacean Codex

"Many years ago on Earth, the rulers of the West wanted a battle to wage. So they commanded their masses, 'okay, you serfs, the surf's up! You're up! The time has come for you to defend our Royal Vanities.' And so the people of You'reUp marched East to fight for the Whole-Leeward Land, and to protect their buddies in beer, their Pals-in-Stein. Many attended the war, and the war did long rage. The Whole-Leeward Land and Pals-in-Stein were destroyed. 'Twas cruel, cruel days, they say'd, very cruel, they all say'd, and thus we record them in the Histories as the Cruel Say'ds. And where was this Evil inflicted? This Evil was reaped upon the realm which is found square in the middle of the *mappamonde*, so we refer to these epochs as Mid-Evil. Afterwards, having been mythed into remission, Prester Prawn plummeted off the Earth which, it is duly writ, had at that time an edge off which to fall. Thusly did Prester Prawn fall to the Primal Pivot, the Center, Zinnia, and environs. Having utterly failed to discover Truth through Earthly exploit and rage, the mighty sovereign now seeks poverty and humility among the Zinnians as sage. A great and noble endeavor, it is chronicled, even if a futile one."

So it seemed that Prester Prawn was, like all kings, an answer to a question that no one had ever really felt it necessary to pose. However, the Zinnians quite liked their Prester Prawn. From what Atollana and Ishmael could determine, he was kindly, helpful, and intelligent. He was also utterly powerless, apparently by choice.

Nonetheless, Ishmael and Atollana both thought that they should now produce the passports given them by Ferdinand and Guinness, this perhaps being the 'diplomatic tangle' of which the sheep dogs had cautioned. But upon seeing the little shabby booklets, Prester Prawn shuddered, and Pliny Plankton nearly laughed.

"Oh, you've been with the sheep dogs, I see!" Pliny said. "I never could get them to understand. This is a *kingdom*, not a *king-dumb*. Bureaucracy has no meaning here in Zinnia."

Zero Degrees On All 'Tudes

With that, Pliny took his quill and added an entry in his Book, reading aloud as he wrote :

> "Let it be further stated that on this day, the seventh day of the seventh moon of the eighty-eighth Zinnian elliptical at the notch of one-eleven in the axis-rotation, it is recorded that Atollana and Ishmael, shrimp scouts with purpose unstated, entered Zinnia and were duly afforded all courtesies possible within the modest means of the Center."

"But where in the Center are we, sir?" asked Atollana.

"In prison," the King answered flatly. "Just as you are on Earth or any place else. We are outside of the tree-lined circle before you. We are, in other words, in prison."

"In prison?" probed Ondew.

"We are indeed in prison," Pliny confirmed. "Perfect beauty flaunts its tranquility before us, yet we are eternally confined to our pitiful incarceration beyond it."

Ondew wondered why they didn't simply enter the circle and not be so abashed about doing so. She offered to accompany them over the tree trunks and into the Center, but received a prompt (though polite) rebuke from Pliny.

"Ondew, Ondew, my dear naïve subject, do you not comprehend this perfect paradox? This place which all lust after *does not exist*. If we were to enter it we, too, would not exist, and would therefore never behold our non-existence. It is the supreme irony of Existence. Not even by the grandest of sacrifices could we experience Nothingness! We are prisoners."

Ondew looked generally confused by all this metaphysical gibberish. "Has, then," she asked, "no one ever entered this circle?"

Prester Prawn looked to Pliny Plankton, the royal chronicler, for the answer. He was actually quite familiar with the answer, as the two had discussed this thoroughly many times, but he wanted the others to hear it from Pliny, the authority on all historical and scientific matters.

"No one," Pliny confirmed. "No one has ever entered within this *Circulus Arbolicus*. The void past these trees is beyond the approach of any and all. We are eternally imprisoned outside it. It is uncreated. It is the fulcrum of that which *is*, and the entirety of that which is *not*. We must live our lives in what little space remains outside it."

This last comment was particularly baffling to the newcomers, as the opening within the circle of bo trees appeared to be quite modest, perhaps only about twenty terra firmate paces across.

"But it is really such a small place," observed Atollana, unsure if she should be self-conscious to be stating the unequivocally obvious. "In fact, I would hazard to say that it is the merest, most insignificant speck compared to the awesomely vast Universe outside it!"

"Nothing but a teasing, heartless illusion of the created world," corrected Pliny Plankton. Forthwith another Zinnian, a macaw person, chanted closer and tried to clear up the newcomers' confusion. He explained to the shrimp and Ondew that

> *"From this side*
> *looking to that side,*
> *little remains*
> *for paint to hide; but*
> > *as that which is nothing*
> > *has dimensions not, so*
> > *what remains is perhaps larger*
> > *than what creation begot."*

Pliny and Prester Prawn both nodded in agreement with the macaw. Ishmael began opening his mouth to dispute the claim, but quickly aborted the effort, for he could find no words to challenge a conclusion derived from such seemingly unflawed logic.

"But this is the center of Zinnia?" he asked instead, trying to simplify the matter.

Pliny and the Prester busily consulted various charts in their volumes and then reiterated, with no trace of uncertainty, that it was indeed the center of Zinnia, as it was the pivot of all things.

"On the other side, where I live," said Ondew, "there is a place similar to this. I think this is just the other end of it. I often stay there. It is unpainted. But there is no reason to be afraid of it."

At this moment a rather helpful looking Zinnian, who had been quietly observing all this strange to-do, ventured forward and used his chant-song to offer an opinion. He was a petite person who resembled the four-legged terra firmate called 'elephant'. His song strongly advised against entering the Void:

"This is where the straight line bends —
At this wall Creation ends.
Though beyond it dry the tears we shed
Past these trees do not be led !"

To prove his message, the little elephant floated cautiously to the edge of the circle, where he extended his long trunk out over some of the lower bo branches so that his song would be funnelled directly into the mysterious Void. When he did this, his song instantly evaporated, falling emptily into it. Nervously, he withdrew his trunk from the disputed place, hesitating for a moment to be sure it had not been injured by the exercise. He then concluded by chanting

"Proof ye have
of what I say.
Enter not this realm,
or ye will dismay !"

His tone was very *serioso*, and he was visibly jittery. Never had he come so close to acoustic apocalypse.

Ondew reached over, gently taking the elephant fellow into her palm as he hovered near the Void. "Your song did not perish," she assured him. "I myself sing within the Void on the other side. It is a quiet song. It is content by itself. It doesn't need to enter the painted world." To calm the elephant's fears, and to settle the question to the little gathering, Ondew secretly decided that she would enter the Center beyond the trees. She first set the elephant person free and then, careful to give no premonition of her plan, jumped between the bo trunks and into the Void. But the moment she did so, to everyone's horror, she disappeared.

All present had to resist the overwhelming impulse to jump in after her. Atollana, who felt a great debt to Ondew, actually began climbing up the ledge formed by the trees, but was stopped by the gathering at large. For all agreed that, whatever had happened to Ondew, it would be senseless to repeat such a disastrous mistake.

A Conference Is Called To Order

In Zinnia, urgent or significant issues are resolved at village conferences. Anyone can call such a meeting simply by alerting the kingdom through his or her song. So a calamity-chant-alarm was sounded immediately, and within moments droves of Zinnians began assembling to discuss a rescue of Ondew. The little elephant, in particular, sang frantically for it.

The first few people to arrive came forward to get more information from the elephant, and in this way Ondew and the shrimp came to know that his name was Tchan.[28] Elephant Tchan quickly chanted to the head of the congress once the chant-alarm was completed. Mr. Mook, the father of squid guide Calamito, also came to the front to help.

[28] The name "Tchan" is believed to derive from the Thai word for 'elephant' (rhymes roughly with 'John'). — ED.

Prester Prawn sat in a chair facing the group. As king, it was his duty to explain the crisis, solicit suggestions, and to arrive at a solution. On either side of him were Atollana, Ishmael, Pliny Plankton, Mr. Mook, and Mr. Tchan (it was really these latter two, rather than Prester Prawn, who seemed best able to expedite the discussion). Sprawling before them were an uncountable array of Zinnians, Zinnians of all sizes and varieties. They gently floated with their own individual song, though all but the person speaking would chant quietly and inwardly, so the whole bunch was hovering quite low to the ground. Anyone who had something to say would float higher with the intensity of his or her chant in order to be heard throughout the assembly. Pliny noted all the discourse in his archives, and all Zinnians pondered each observation and suggestion with equal seriousness.

A sea horse suggested that Ondew had slipped straight through to the other side of the Center, thereby ending up back on her own side of the Middle. The sea horse commented that when reaching the surface on the other side, the terra firmate woman was probably travelling at such speed that she shot up into the sky a bit before coming to rest on the ground. He added that in all likelihood she came to a reasonably soft landing in a basil grove. Or, perhaps, she could have landed in a rice paddy, since relatively random trajectories might favor such terrain. In addition, the woman's internal chant might have caused her to lean one way or another, adding yet another variable to the complex equation. In the final analysis, he consented, it was difficult to be sure.

An owl confessed to the fear that Ondew had perished, completely and irreversibly. It was a very moving and despairing chant, in a minor key with lots and lots of flats; but perhaps most disturbing of all was the song's ominous ending with a quiet plagal cadence. The entire congress shivered.[29]

---

[29] In music, a plagal cadence is a cadence from IV to I (rather than V to I). It is often found in early Church music, and is typically used with the word 'amen'.
— ED.

One of the more curious remarks came from an elderly caribou. It was her opinion, she explained in her chant, that we had all merely failed to understand what someone would look like to us while in the Center. It was her belief that space and time are products of the created world, and that therefore neither exists within the Middle, which was, of course, unpainted. As a result, according to this hypothesis, we would not see the terra firmate girl, nor, perhaps, she us. Nor would she realize that she was being missed, for if time were strictly a dimension of the painted world she would not be aware of the length of time that had transpired. This made absolutely no sense to most other Zinnians, although her chant was quite nice, with a tangy Mixolydian twist to it. Some Zinnians thought that perhaps the old caribou had spent too much time in the aromatic Aeolian groves. However, a monkfish, who seemed more receptive to the caribou's comments, suggested that perhaps meditation might be the path toward understanding the Center and the terra firmate woman's fate.

It was then that Ishmael first noticed a cloaked, rather grayish figure slowly begin to emerge from the distant, misty blur of Zinnia. This person was a two-legged terra firmate, and like the other terra firmates she was unchanted, that is to say, she walked on the ground. She was in fact quite like Ondew, except that she was of a fairer complexion, with golden, straw-colored hair. And, she wore clothing; a musty toga with a hood covered her body and scalp. Ishmael hoped that clothing was not symbolic of the uncivilized sort of terra firmate that he so feared back on Earth. But, then, Pliny Plankton and Prester Prawn wore clothing, and they were both peace-loving creatures indeed.

As the new terra firmate woman slowly walked toward the middle of the congregation of Zinnians, all the varied creatures in her path moved over slightly to allow her through. Once at the hub of the crowd, which of course was still a fair distance from Ishmael and Atollana, she knelt down and placed her hands on her knees (by kneeling down she put herself on a better level from which to address the Zinnians, who in their general hush were now hovering only about knee high to her). For a few moments there was silence; everyone waited for her to speak. Whoever this woman was, the Zinnians obviously regarded her with special reverence.

"My dear friends, I have heard the disturbing chant," the terra firmate began. "None of us can surely know what has happened to the traveller from beyond Zinnia. We know that she lives on the other side of the Center — that is, her land is antipodean to ours. Her feet and our feet join at the *terra*. Perhaps she is unharmed within the Center, or has slipped through back to her side. Or perhaps she has perished in Nothingness."

She paused briefly, took a deep breath, then said "But I feel that — both in the attempt of a rescue, and for reasons of my own self-destiny — I must follow her into the Center myself."

The people of Zinnia were flabbergasted at this announcement, and a nervous murmur sped through the crowd. From their murmur, of course, they suddenly hovered further off the ground, with some of the more emotional Zinnians briefly fluttering above the girl's head. Ishmael and Atollana did not understand what was happening, so Mr. Tchan quietly explained.

In a whisper he chanted :

"She who speaks is Honua, Custodian of Grief. All that forsakes Joy is her burden. She lives as a recluse in a cave atop a mountain."

The two shrimp immediately recognized the name; she was the terra firmate girl who had sculpted the poem on the shell. Atollana held up her necklace to Mr. Tchan and confided to him its verse; the elephant, stupefied, softly nodded. The villagers, meanwhile, were trying to dissuade this Honua from her mad proposal. Sundry chanting voices begged her to forget the reckless scheme :

"Surely we'll lose you too!"

"You'll be gobbled up by perfect Nothingness!"

"Please don't martyr yourself!"

"Too much haste for such a grave decision!"

"No one can enter the Center!"

... and on, a scrambled dissonance of pleas. But she would not abandon her plan.

"My dear people of Zinnia," she continued, "let me explain my proposal, for I will need your help. As you know, by my cave there grows a great and ancient vine-plant, with arms long enough to extend far past the Center, if unravelled. We will sort four arms of the vine. I have already discussed this with the plant, and he is anxious to help. First we take two of the vines and securely tie one around each of my shoulders. The other two I will tie around my waist. Now, as it is important that you remember which pair is which, we will use reddish ones for my shoulders, and greenish ones for my waist.

"You will then lower me into the center nearly as far as the vines will extend. Leave just a bit of slack behind you. At this point, four squid-people will anchor the vines, staying close above the center but being very careful not to approach too close to Nothingness. Each of the four vines needs a squid to secure it in his or her tentacles. Change squid-shifts often, so that none of you gets too fatigued.

"Now listen very carefully. If I tug on either of the vines tied to my shoulders — the reddish ones — you should pull me up out of the Center. If I tug on either of the vines tied about my waist — the greenish ones — you will need to summon two elephant folk. The two squid holding the vines which are tied around my waist should then move up slightly to allow each elephant to get in front and coil his or her trunk around the vine. Once both elephants have a firm contact, they will begin a chant, which I will explain. The vibration will travel down the vine to my body. Their chant will be in rhythmic unison, but precisely a tri-tone apart in pitch."[30]

Zero Degrees On All 'Tudes

This very last instruction clearly shocked the Zinnians once again, for they all shuddered and momentarily soared in their chant-hover. One particularly distressed Zinnian, a trumpetfish person, shot up so quickly that he choked from the jolt, crashing into the ground briefly before regaining a normal minimal chant-height. Everyone knew the trumpetfish was high-strung, and those nearby rushed over to make sure he was alright.

Ishmael asked Mr. Tchan to explain what a tri-tone is, and why it so upset the crowd.

"When the natural whole, the perfect octave of peace, is ripped in half, it becomes two very unhappy half-wholes," the elephant explained. "These are tri-tones. Tri-tones incessantly plead to be made whole, that is, to be re-joined with their other tri-tone and thus form an octave once again. They are called tri-tones because they are made from three whole steps. It must be three, because only a triangle could withstand such unstable gravity.

"For this reason the Lydian ferns are revered with an almost mystical awe.[31] Their very psyche is predicated upon a tri-tone — you see, the fourth of their scale is raised, forming a tri-tone with their own tonic. *Yet they are esteemed for their wisdom and introspective peace.* It is a transcendent contradiction, a true miracle of nature. Thus from Lydian ferns one might learn to find peace despite sorrow or hardship."

---

[30] In music, the tri-tone is a dissonant, highly unstable interval named for its three whole steps, equalling exactly one half of the octave. The same as an augmented fourth or diminished fifth. See also note below regarding the Lydian mode. — ED.

[31] In music, Lydian is the mode whose tonic would be 'f' with natural pitches. Of conventional Western modes, the Lydian is unique in that a step of the scale (the fourth, e.g., the note *b* if in the key of *f*) forms a tri-tone (being in this case an augmented fourth) with the tonic. Beethoven used the Lydian mode in the slow movement of his opus 132 string quartet. The phenomenon of a tri tone with the tonic occurs in other forms as well, such as in the whole-tone scale, some Asian music, and Western music influenced by folk or Eastern music; Béla Bartók was particularly fond of it. — ED.

The Crustacean Codex

So it seemed that since the Lydian mode has a tri-tone — that is, a dissonance, a *pain* — against its tonic — that is, against *itself* — Zinnians regarded the Lydian ferns as having come to terms with the great contradiction of life and suffering. Mr. Tchan stopped to glance around the assembly, but saw that not everyone was calmed yet. He settled back and resumed his impromptu primer on tri-tones.

"There are two basic ways that tri-tones are interpreted," the elephant continued. "All creatures are born a tri-tone, say some. That's why people run about to find their other half, and when they do, they hug and squeeze each other until making themselves one, that is, into a whole octave again. The other philosophy is also very beautiful."

"Yes," carried on Mr. Mook. "Others chant that we are *not* born tri-tones. According to this thinking, we enter Existence as complete octaves or unisons, and that if two people love each other they vibrate sympathetically. The concept is that we get torn into tri-tones only if we are not at peace with ourselves."

"Both chants are very pretty," commented Mr. Tchan, "and certainly they are both true. They are the complementary forces that move two people to love each other."

"But as for singing a tri-tone," stressed Mr. Mook, "to sing a tri-tone deprived of its other half is a painful thought indeed."

"So people never *want* to sing in tri-tones?" checked Ishmael.

"Never, never," answered the elephant, both he and Mr. Mook aghast at the thought. "Not parallel, consecutive tri-tones, anyway. Well, of course, I do remember one year at Fugafest, the Phrygian Grove scandalized the night with an inverse cannon in tri-tones. Many of us overheard it. Chilling indeed! But meant in jolly fun."[32]

---

[32] In music, Phrygian is the mode whose tonic would be 'e' with natural pitches. An example of this mode is the sublime second movement of the fourth symphony of Brahms. — ED.

The general squeamishness caused by Honua's last instruction had finally eased, and everyone's attention refocused on the congress. Honua attempted to explain her strategy. The exact instructions were most important :

"The reddish vines are to assist in the event it is difficult for me to pull myself out of the Center. The greenish ones are to transmit the tri-tone chant through the stillness of Nothingness. And the tri-tone chant itself is to pluck me from the timelessness of the Center. Assuming that I do not perish in the Center, I may be quite unaware of the length of time that has passed.

"For this reason," she continued with one final point, "if on the seventh day you have not gotten any signal from me, you must then execute the tri-tone chant on your own."

Honua finished her speech by asking Prester Prawn to coordinate the plan. Pliny Plankton, who had been busy recording the entire congress in the Archives, ended the entry by adding :

> ' . . . so thus did the People of the Kingdom of Aqua-Abyss in Zinnia on this day scheme to rescue their guest, known as Ondew, from the Unknown of Nothingness. And thus will Honua consummate her destiny, that is, finally and at last, to heed the fate that awaits her in that place where perfect peace remains within'.

With great urgency the assorted creatures of Zinnia all bobbed over to the Center. There they found that the great vine had already neatly laid out four of its monumental arms, having carefully chosen its two reddish and two greenish ones best suited for the task of Honua's descent into Nothingness. Great length, of course, was necessary for all four; but whereas strength was a primary virtue for the reddish ones, the greenish pair needed to be more supple and tender in order to carry the vibrations of the elephants' chant with greater fidelity. Prester Prawn proceeded to the edge of the Center to coordinate the steps of the plan, while squid grouped themselves into several shifts of four individuals each. In addition, one extra squid would always stand by to take over from any who became fatigued or ill. Off to one side, little hovering elephants were busy practicing the prescribed tri-tone chant. It was an abrasive exercise.

Honua, in the meantime, had quietly run off to bathe in a stream which bobbled near the mountain where she lived. The Center was as venerated as it was feared, and she did not want to pollute it with a grimy body. Bathing was also something of a sacrament for her, as she did not know what would happen to her once she lowered herself within the circle of bo trees. Perhaps the most pessimistic of Zinnian theories about this was correct: perhaps she would simply cease to exist.

While she was bathing, Calamito, the squid youth, timidly approached the stream, pausing where Honua had laid down her robe. He was holding an object in his tentacles. Honua noticed the visitor and motioned him to come closer (the stream was thrushing a bit too much for them to have spoken from a distance).

"Forgive my disturbing you, oh Honua," Calamito began as he ventured nearer. "The Elders believed that this might be important. The shrimp damsel who came to Zinnia with Ondew wore a necklace which had been given to her by an Elder back on her plant Earth. When you arrived at the village conference, she recognized your name as being the same that is on the necklace. She informed Pliny Plankton and asked that perhaps it could be returned to you. The Elders all thought it a good idea. And the Shrimpess asked — asked me to express her love to you."

Honua, now sitting on a large rock in the stream, took the necklace from the little squid's arms and fixed her eyes on the shell and its poem. Slowly pulling her eyes from it, she smiled at Calamito and thanked him for his kindness. She then placed her head on her knees and quietly wept. The little squid, without saying anything more, slipped away back to the Center to continue preparations for the rescue.

Honua left the stream and put on a fresh robe. It was white, and in the shape of a tunic. She took the necklace into her cave and, without ceremony, gently placed it on the ground in the corner where she slept. The girl then left the cave and set out for the Center, perhaps, she thought, never to return.

Following the great vine's long arms that had been selected for the task of entering Nothingness, she reached the crowd of Zinnians hovering about the ring of trees surrounding the Center. Everyone hushed. Prester Prawn came up to her and asked if she might reconsider her desperate mission. She would not.

To avoid the risk of sentimentality, Honua hurried toward the ominous circle of trees, pausing only to afford herself a final brief glimpse of the land and creatures of Zinnia she so dearly loved. Attendant squid brought the ends of the pair of reddish vines, which they secured around her shoulders, and the pair of greenish vines, which she clinched around her body in the fashion of a waistband for her tunic. Without pomp she climbed over the bo trunks and motioned the squid to lower her into the Center. Her body disappeared as it passed the ground level. Last to disappear were her arms and hands, which clutched the two reddish vines above her head.

Thus was her modest rite of passage into Nothingness.

Although she did vanish as feared, an aforementioned detail offered the Zinnians reason for hope: as her body slipped unseen below ground level, the part remaining above appeared unharmed. Her face, and indeed her expression, betrayed no distress. Perhaps the old caribou was correct; perhaps it is only time and space that cease to exist within the Center.

The Seventh Day

Six long days came and passed without any signal from Honua. No one in Zinnia had been able to continue life as usual during these days, such was the worry over the two women who had vanished into the mysterious void within the ring of bo trees.

We should have boarded close the void long ago, precisely to prevent such a catastrophe, chanted one Zinnian, with a terrible sense of guilt.

No, no, reassured another, *we could not have done that, for to do so would have been to deny everyone free will.*

Well, we should at least have tried harder to discourage Honua from sacrificing herself, lamented others.

No, insisted the more philosophical hold-outs in the crowd, *we made our concerns clear to her, and she choose to proceed. It was not a rash decision, but rather one which was the culmination of her long years in Zinnia. It was her desire.*

In the end, all Zinnians concurred on one point: The fate which had befallen Ondew and Honua, whatever its nature, was that which they chose. They may well have both made a *mistake*, yet not necessarily a *wrong judgement*; there was an important distinction between the two. And still, perhaps naïvely, many hoped for their safe return by the seventh day.

On the seventh day, with still no signal from Honua, two well-rehearsed little hovering elephants took their positions by the two greenish vines, ready to initiate the tri-tone chant. The squid grasping those vines moved back a bit to allow the elephants to coil their trunks around them without getting too close to the Center.

Both elephants in the duo had been quite strict with themselves about rehearsing the chant frequently so as to maintain the tri-tone's potency. Without such practice it risked slacking down or up to a uselessly stable fourth or fifth. The time having finally come to perform, they looked at each other

Zero Degrees On All 'Tudes

to cue the chant, and sang in meticulous, excruciating parallel tri-tones.

Immediately upon the precipitous climax of the strange duet, a sister of one of the elephants rushed over to massage their trunks to avoid any danger of temporary tri-tone trauma. Her best friend, a squid who was the sister of one of the vine pullers, also hurried there to assist in the preventive therapy, but had to carefully weave her way around the Center to avoid getting in the way of the various goings-on by the vines.

As the general gathering of Zinnians thanked the little elephants for their efforts, all were surprised to hear a soft drone in the distance. The Gregorian Grove Elders were intoning an exact inversion of the elephant's tri-tone chant, but in true fifths and elongated to half the speed; with this peaceful gesture, the Gregorian Grove was restoring the low-energy serenity to which Zinnians were accustomed, purifying the fiber of the Kingdom from the tension of the tri-tone chant, just as it cleansed it of shakiness after each Fugafest. This, of course, further calmed the two brave elephants.

While the Gregorian Grove was still busy with its reciprocal chant, anxious Zinnians silently huddled around the ring of trees hoping for some signal from the void within. There was none. But just as the squid, in desperation, were about to pull up the four vines, a figure reached a hand up from within Nothingness. It was Honua. She clutched onto a bo trunk, and pulled herself out from the Center, two greenish vines still secured around her waist.

Honua smiled at her countrymen, then motioned the squid to resume towing the two reddish vines. Several long moments later Ondew emerged from within the Center, the ends of the vine's great arms secured to her shoulders and chest. As ecstatic chants whirled about the Center, Ondew and Honua saddled themselves on a thick, low, and nearly horizontal branch of one of the bo trees, their feet dangling above the freshly-exonerated Void. All of Zinnia emotionally crowded around the two women who had returned from the Unknown. Everyone hushed.

(Their hushness, however, sent them floating too close to the ground to see the terra firmate women well, so many resumed a silent high-energy chant — similar to a tense silence in powerful music — to maintain a high yet perfectly quiet hover. Those that did were the first to notice that Honua seemed to be smiling. No one in Zinnia had ever seen her at peace.)

Honua surveyed the crowd and spoke :

"Thank you, dear friends, for your help. As you can see, we are both unharmed by our visit to the Nothingness beyond the circle of trees. I had no idea that several days had gone by, for — as some of you suspected — time and space apparently exist only in the painted world. That, of course, is why we prepared the tri-tone chant. The sound travelled through the vines and jarred me from the tranquility of the Center."

Honua turned to face the little elephants resting from the tri-tone chant, placed her open hands together, the tips of her fingers just below her mouth, and gratefully lowered her head. The two elephants blushed and showed their trunks to assure her that they had developed only minor blisters from the experience. Honua nodded, and turned to face the crowd once again :

"I must explain that what Ondew had told us is correct. The Center extends straight through the *terra* to her isle, which lies far below our feet. Although this was the first time she entered Nothingness from the Zinnian side, she retreats to it often from her land. Now, I will tell you that Ondew was the only one who had ever entered Nothingness — and that in a sense this still remains true."

Suddenly the crowd's silent chant broke into a low murmur as Zinnians tried to reconcile this paradox. How could this be? How could Ondew be the only one ever to have entered Nothingness, when both she and Honua had just returned from there? Honua explained :

"Ondew knew Nothingness before she knew the painted world. I am a variation of Ondew, conceived with original Creation. I am her twin sister. I am, so to speak, the other side of her reflection. We were separated when Creation was born. She cultivated her isle-cosm, while I grew with the world beyond."

At this revelation, the more intellectual among the crowd observed that Ondew and Honua are the Sacred and Profane echoes of each other. Honua smiled at this interpretation which, she thought, was really quite insightful. When life got painted, pain, suffering, sorrow and despair got painted with it; long ago she had placed these contradictions upon her own shoulders, a responsibility which her beloved sister Ondew had never intended. But Honua, like all life, painted herself according to her own visions once she had been created. Thus as the manifestation of the Profane, she wore clothing; her sister, the embodiment of the Sacred, did not. She bade them listen further for one last untangling thought :

"Wait, dear friends, there is more I want to tell you. *I wandered here where none ask why, and set upon my tears to dry.* You see, in the stillness of the Center I came to peace with myself and with Creation."

That, she thought, explained it better.

VIII
Fugafest

Honua smiled shyly and took Ondew's hand. She led her off the bo limb, onto the tree trunk rim, and finally onto the ground.

"Excuse us for now," she apologized as they walked toward the Chanting Jungle, "for within me are songs I have long wanted to sing to my twin."

Zinnia beamed with happiness and relief. Pliny Plankton busily recorded the event and made several notations to himself to amend the Histories regarding Nothingness and the Center, which were obviously not entirely accurate in stating that no one had ever entered within it. Rather, only the original inhabitant of Nothingness — and her twin, her painted worldly incarnation — had been there. Though his original Historie was correct in essence, it was clearly in need of a very important footnote. And despite the revelations about the Center, all Zinnians seemed resolute that they would never enter it themselves, as though it possessed a sanctity not to be disturbed.

As Ondew and Honua walked away from the gathering, Prester Prawn, sovereign of Zinnia by default and anxious to do the best for his subjects, broke the respectful quiet with an announcement:

"A celebration will be called!" he proclaimed, "let's celebrate their safe return and the wisdoms they have brought!"

Fugafest

Pliny Plankton, noticing an annotation in his Archives, leaned over to him and whispered, "you may not need a celebration, dear chap — according to my calculations, tonight begins the Fugafest."

Pliny Plankton's charts were accurate. Shortly they began to sense faint quivers in the ground. The Universe was approaching its precise alignment with the Center, and powerful interplays of gravity were beginning to tug nigh and about, creating small benign tremors. Far from Zinnia, Yandu was getting his annual cosmic massage, and Ferdinand and Guinness were busy patrolling the Yanduian Highlands with a quake alert.

With these tremors, pairs of very, very low silvery, guttural voices began to sing the three syllables of 'Fugafest'. They sang in parallel fifths, descending a minor third for the last syllable. Each pair of voices sang independently, not being coordinated in tempo or pitch to the other parallel fifth pairs. Further, the pairs of voices were often separated by great distances, their chant coming from many corners of the Kingdom.

"Frogs," explained Prester Prawn.

"Frogs?" asked Atollana.

"Frogs, both the lowest bass and baritone frogs," confirmed Pliny Plankton.

Soon the voices of diverse creatures intoned many other images on *fugafest* motifs, and the collective music took on the life of an amorphous breeze. A rather studious fern who wished to preserve the wondrous chant unrolled a parchment sheet specially saved for this day and neatly drew the five straight, parallel lines of a music staff. While notating the sounds, however, the energy of this invisible music-wind began contorting her staff, so that the lines took on the shape of a water sprite gliding through an ocean river :

"Fuga_fest... Fuga_fest... Fuga_fest..."

The delighted fern affectionately asked the naiad to be careful with her legs, lest she inadvertently knock any notes to the wrong pitches. But at this attention the staff lines straightened with a self-conscious blush and a nymphish wind whoozed from the parchment. Though the fern regretted frightening the sprite, she was buoyed to have had their brief encounter, and gleefully passed the remainder of Fugafest notating whatever music was within earshot.

The shimmering cacophony of Fugafest proclamations gradually increased, soon completely permeating Zinnia, then slowly diminished and subsided. As it eased, an occasional plant giggle could be heard. Fugafest had begun.

An early physical sign of Fugafest was the misty fog which seeped from fissures in the Zinnian landscape, apparently the result of ground faults which loosened during this period of geological strain. The steamy cloudiness was part of the magic of Fugafest, for it transformed Zinnia into a surreal realm where visions passionate and fantastic could flourish, and it artistically camouflaged fugallers who wished to remain discreet. Soon, in various and brilliantly improvised maneuvers, some of these troubadours began to exploit its foggy cover.

The silhouette of a Mixolydian fern was visible racing toward the Dorian grove. But in his haste to be discreet, he crashed into a Phrygian fern who herself was secretly stealing over toward the Ionian Lowlands.[33] Each laughed and tidied up the other, and after mutual apologies both scurried off once again, embarrassed at having been found out. The Mixolydian fern's

Fugafest

rendezvous, it turned out, was with the same Phrygian fern's sister. These two slipped down the banks of a local stream where they sang a very simple, graceful fugue, and became sweethearts. Each had, for a long time, secretly been enamored of the other, and Fugafest presented them an opportunity to overcome their shyness.

Although the bank of the stream was only moderately high, it was fairly steep, and there was only one narrow passageway to it for quite some distance. Fog accumulated by the path to guard the lovers once they had passed; when they reached the edge of the stream, they were fully hidden below its bank. Clothed in Fugafest mist, which had risen from the pores of the very *terra* itself, the two fugue-voices clutched each other, mesmerized by the rushing water at their feet. Utterly at peace with Creation and with each other, both unknowingly slipped into a light slumber. But so harmoniously did they vibrate together, that upon awaking there was a passing moment in which neither could remember which of the two they were, for indeed in their dreams they were both.

The two were now equal fugal subject and counter-subject; that is to say, they were the most beautiful of lovers.

Fully awake once again, the Phrygian maiden filled her chest with modest breaths and puffed away the *terra* fog-mist which cloaked them, first baring her lover's body, then her own. The mist, skirting the shallow part of the stream and its perimeter, now formed an indistinct wall to cloister the two from the rest of the Universe, being careful not to intrude upon them again until they were ready to rejoin it.

Soon after these two had first settled by the side of the stream, the other Phrygian sister (the one who had earlier collided with the Mixolydian boy) had reached her appointed place in the woods up in the nearby hills. Now, this particular

---

[33] In music, Ionian is the mode whose tonic would be 'c' with natural pitches (i.e., a 'major' scale) Curiously, this mode, though the most common in traditional Western music, is said to have been considered a 'wanton' mode during the Middle Ages, suitable only for secular music. — ED.

The Crustacean Codex

Phrygian — a Micronesian Phrygian — was the fancy of many earnest youths, several of whom hoped to win her favor this Fugafest. One of these suitors was indeed dear to her heart, but no one, not even he, knew which of them it was, for she had always been very private with her feelings. Nor, as they were scattered among the misty, wooded hills, was it possible for her to see any of them. So, in a tradition perpetuated by Micronesian ferns since ancient times, the girl sang a simple fugal motif and listened as each of her suitors introduced and developed a second statement of the fugue. Each had his own peculiar, distinguishing touch for this, of course, and from this fugal 'signature' the heroine easily identified her chosen lover. Her fugue, in turn, responded only to his, and so the lucky lad knew that it was he whom she adored.[34] Their mutual love was now confessed. The two searched for each other by following the sound of their chant, for they were veiled by haze until reaching within a mere few steps of each other. Finally within sight, only a scant further moment was suffered before the two at last touched, impetuously melting their consecrated voices together into a unison.

There passed a moment of eternity.

Time then momentarily suspended itself as the girl gently eased their embrace so that she could lay herself against the earth. Her body now united with the *terra*, eternity was resumed. The two caressed their innermost chant around each other, and thus these two beautiful ferns became a beautiful fugue. Their counterpoint was brimming with rhythmic energy, joining in powerful and euphoric harmonies until, in one final and impassioned youthful cadence, together they

---

[34] The only identifiable parallel to this motif-identity custom among lovers is found in some Micronesian cultures. At night, a suitor would insert a 'love stick', a wooden stick with carved designs unique to that particular boy, through the thatching of the hut wall of the girl he desires. Upon hearing the stick pierce the wall, the girl would feel its carved motifs, in this way being able to identify the suitor. If the carved designs are those of the boy she wants, she will pull the stick into her hut and the boy will know to slip inside to visit her; if not, she will push the stick back out. — ED.

shared their very life's energy. Slowly, they slumped blissfully into the mist-cloaked mountain grove.

> It is a most joyful of natural mysteries that
> in this ultimate act of life,
> two souls join to become one, a perfect octave,
> yet both always remain whom they are.

The girl's other suitors, of course, knew that their advances had been rejected because her counterpoint had not responded to the particular designs and nuances of their own fugue. Though disappointed, they nonetheless felt a warm happiness for these two fugue-chants who, on this lovely day, had found their counter-statements, their other halves. In a sense, they were all part of the eternal fugue of life.

Far off in a quiet gully by a secluded grotto, an Aeolian couple slipped through a Molucca thicket to a hidden garden where, many years ago, they had first met. The two sang carefree *canto* and recitative, both of them modestly veiling their fugal motifs in pretty filigree and flourishes (some of which became little cadenzas). After a peaceful respite the soprano Aeolian, singing an unaccompanied passage, stood before her lover and allowed these simple embellishments to drop quietly to the *terra*. She shed every lace and frill, leaving her very chant in its most simple and perfect beauty, unclad before Everythingness. Silk-like clouds of gentle tremor mist blessed the garden and their bodies.

For a while the fern chap loosely improvised above the girl's continuo, which was a soft chant-figure in parallel fifths.[35] But soon, beckoned by her penetrating chant-figure, he caressed his counter-motif between her air's open intervals. The two embraced in a most splendid fugue, one which, in unending variation, they had known since their youth.

---

[35] A "fifth" in music is the interval produced by the first and fifth steps of the scale; in all conventional Western modes this yields a 'perfect' fifth, acoustically the most stable interval after the octave. When a fifth exists by itself, lacking the third or other interval, it is referred to as an 'open' fifth. — ED.

This was, of course, the less sensational and rarely noticed side of Fugafest. But if not the stuff of juicy legends, their fugue chanted in ecstasy no less sublime than that of the young lovers on the wooded hill. Fugafest offered all Zinnians a chance to leave their daily chores and be with the other voices of their fugue.

These are tastes of the many episodes that made up Fugafest that year, the Fugafest itself being a massive polyphony of joyful stories. There were endless such adventures blooming all over Zinnia. But despite the Zinnian appetite for tales of Fugafest, few ever became etched in Zinnian lore, because few had any witnesses within earshot except fellow fugallers far too preoccupied with their own counterpoint to take note. The rich Zinnian repertoire of Fugafest stories was, rather, the fruit of the extraordinary ancientness of the tradition.

The only other fragments known from this particular Fugafest are of a more capricious spirit. A Lydian flower was sighted just outside one small village, ducking into nooks and corners as she advanced, surveying the area carefully before each flight to her next hiding spot. She seemed to be heading toward the Aeolian grove, but her constant detours made it difficult to be sure. When teased by curious friends the following day, the shy Lydian smiled and revealed only that she had gone swimming in a nearby lagoon with someone who adored playing sonatas with her.

Quite out in the open, a teasing ballad between two succulents, a bashful chromatic aloe and a coy whole-tone cactus, was plainly heard. The cactus was coaxing the aloe to abandon her chromaticism, which she was reluctant to do at first. But when the cactus began imitating her by bending his whole-tones, and then used selected whole-tones to imitate a passing ginger root (who was pentatonic), the aloe starting laughing and attempted a retrograde variation of it.[36] This proved to be overly ambitious, however, resulting in several episodal crashes, but a kindly Dorian vine chanted over to offer her admiration for their effort.

Fugafest

A different sort of escapade was organized by a consort of water reeds. These reeds fastidiously arranged themselves into an untempered scale, each reed handling one pitch exclusively.[37] The roguery began when their fugue modulated to distant keys for which their untempered scale was singularly unsuited, and each reed had to tediously concentrate to keep from either adjusting their pitches to the new keys, or slipping into a comfortable all-purpose tempered scale.

It happened that one viola reed who happened to be the third of the original minor key found himself being the seventh of a major key on the lowered fourth of the original key. The pitch this reed was obliged to produce was so excruciatingly low for its new role that a gamelan vine (who happened to be passing by) gave the reed a friendly squeeze to 'up' his pitch a bit and relieve the poor fellow's discomfort. Although the fugue continued despite the other reeds' near disintegration into uncontrollable laughter, the urge to return to the original key now became so obsessive that the fugue took a series of abrupt and decidedly inelegant modulations until, at last, they were home again.

The Earth Plant Cries Out

So the Fugafest progressed, gradually gaining momentum until the multitude of frenzied fugallers was so overwhelming that some counter-subjects and episodes became hopelessly

---

[36] *Chromatic* refers to successive half-steps; thus a chromatic scale is based on twelve notes, rather than the seven of a conventional diatonic scale. *Pentatonic* refers to music based on a five-note scale; the octave can be divided into five in many different ways, some equidistant (found in Southeast Asia and parts of Africa), some with highly unequal intervals (China and Japan). Among the Western composers who were fond of pentatonicism were Debussy, Ravel, and Puccini. — ED.

[37] 'Tempered' refers to the system of musical pitch in which the true, theoretical acoustic steps have been compromised ('tempered') so that all steps are evenly spaced. Transpositions or modulations on keyboard or other fixed-pitch instruments are essentially impossible without this system, hence, apparently, the origin of the sport undertaken by these reeds. Tempered tuning is in virtually universal use in the West, and common throughout the world. — ED.

tangled, and various counter-expositions were inadvertently picked up by the wrong fugue partner (a legendary source of Fugafest lore).

When the day of fugal feasting came to a close, the Gregorian Grove began chanting the same 'fugafest' motif that the deep-throated frogs had intoned at the beginning. This served a purely poetic purpose; even those with the sturdiest fugal stamina were by this time well in need of a good sleep, and the Gregorian Grove Elders' peaceful chant provided Fugafest a tidy, yet never strict, ending.

As the Zinnian landscape calmed and vaporous clouds melted back into the *terra*, two pairs of figures descended from twin promontories and chanced to meet. They were Ondew and Honua, and Atollana and Ishmael, all of whom had quietly observed Fugafest from cloistered hilly perches. The four climbed a wide, rather horizontal tree limb that nearly spanned a brook which braided the valley between the two hills. Each found a comfortable nook on the tree branch and, offering themselves to the damp, slightly brisk quietude of the Zinnian night, gently drifted to sleep. Fugafest was over.

To Ondew and Honua, the symbolism of Fugafest was as evident as it was natural. Fugafest was a joyous exaltation of the Universe's being infinitely diverse, yet ultimately one. The voices of a fugue are at once identical and different, at once contrary and complementary, at once one and many. A "voice" was of course a person, but it was also a rock, a wind, a galaxy, or the moonlight piercing through a distant cloud. Every element of Creation was a voice of a fugue. Zinnians understood the poetic meaning of Fugafest, and though there was rarely a need to intellectualize it, they knew that everyone was born a new voice in the great fugue which was Creation. And they also knew that Fugafest coincided with the day Zinnia was aligned with the axis of the Universe.

But the connection between these may not have been purely poetic. The Elders of the Gregorian Grove, as well as some of the more academic-minded Zinnians, had long suspected that the ancient custom had also evolved as the Zinnians' means of distracting themselves from their annual day of unnerving tremors. Fugafest was comforting because it treated the trembling of the *terra* as just another voice of the Great Fugue. This year, however, the value of Fugafest to help Zinnia cope with the tremors became quite evident, because this year, after Fugafest was over, after all of Zinnia was peacefully asleep, the tremors began once again. Never had that happened before, never was such an occurrence even alluded to in ancient chronicles. The Kingdom of Zinnia was suddenly afraid.

Although these new tremors were strong enough to awaken everyone in Zinnia, at first many believed that they were merely freak after-shocks. But the quakes worsened, and an emergency meeting was called. All Zinnians who were able, however fatigued from Fugafest, attended.

Pliny Plankton, for formality, immediately entered the conference into the Histories. He then consulted sundry astronomical charts in the Archives and determined that the annual axis alignment of the Primal Pivot and the Center could account only for the brief, benign tremors of the Fugafest day. These new, more severe tremors could not be attributed to that momentary cosmic phenomenon. Making a series of calculations with abacus, he isolated the one body in the Universe whose precise position at that moment could affect Zinnia in such a manner. It was Earth.

As the Zinnians hurriedly gathered, Pliny Plankton briefed Prester Prawn on his cosmological findings. Both agreed that they should quickly inform Atollana and Ishmael of the news about Earth before the session began. Needless to say, the two shrimp were stunned. Prester Prawn requested that they speak to the Zinnians if questions arose.

Everyone was now assembled and eager for Prester Prawn to explain what was known about the situation. He possessed a far better understanding of the *Extrum Zinnian* and *Extrum Yanduian* world than did the Zinnians, and that virtue was particularly appreciated now.

"My beloved subjects," he began, knowing that this might be the most important delivery of his presterhood, "these tremors are not caused by the cosmic alignments for which we celebrate the Fugafest. Rather, these are hurtings somewhere in the Universe. Some organ of Creation is in pain. And we believe that organ to be the Earth Plant."

Zinnia echoed motionless as Pliny Plankton's findings were pronounced and explained. The shocking revelations made little sense to anyone, though most understood that Earth was the particular limb of the Universe they had come to know from Atollana and Ishmael. When the Prester concluded his summary of Pliny's assessments, he turned to Pliny himself who, as the most learned in matters of cosmology, would now take over the meeting.

All attention shifted to Zinnia's resident scholar of natural science. Pliny told the crowd that he would supply facts, that most cardinal pillar of reason, as best he could, but that he needed their help to mold those facts into a remedy.

One Zinnian, a puffin, chanted a bit higher and asked that the precise ailments of the Earth plant be explained :

> *"The plant called Earth*
> *place of our guests' birth,*
> *we know they seek cures*
> *for a disease which it endures.*
> > *But perhaps we could better determine*
> > *what might quell our quake alerts*
> > *if the Earthly shrimp could now sermon*
> > *as to why the Earth limb hurts."*

A sound and sensible request, all agreed, and so the shrimp addressed the congress. They told the bewildered people of Zinnia of distrust between people who had life figured out differently, enmity between people because of their particular chant, wickedness against people because of their particular mode. They told of malevolent creatures who posed as presters, controlling the multitudes to their own gluttonous ends by teaching them chants of hatred and ignorance. Lastly, and perhaps most shocking, they told of the utter devastation of the Earth plant, of the deliberate poisoning and ravaging of the entire Earth voice of the Universe.

Zinnians hovered motionless in horror as they listened to grisly tales of entire chanting jungles being systematically slaughtered, of mysterious and sometimes invisible poisons being reaped upon the Earth plant. They were told that substances of death were being funnelled into the seas and pumped into the air, and that the entire gorgeous Earth orb, along with its precious opulence of life, was exterminating itself.

It would be too painful to elaborate upon their reaction to the news of this havoc, but it will suffice to report that never has a sadder chant been heard than that which permeated Zinnia this day. There were many *portamenti* to be heard, excruciatingly painful *appoggiaturas*, as well as lowered minor sixths of the scale and many, many other tearful -isms.[38] Even the comforting Picardy Grove, the Zinnian bastion of auspicious endings, could not anoint this sadness with any warmth. No one in Zinnia had ever known the Picardy Grove to let a cadence languish in such torment.[39] Far in the back of the conference, one Zinnian who had momentarily left to tend to an infant (who was frightened by the tremors) had just returned. His neighbor tried to relate to him what had transpired :

---

[38] *Portamento* is from the Italian *portare*, literally, "to carry." In music, portamento is the connecting of two melodic notes with a "slide" rather than a distinct change to the new pitch. The full duration of the portamento almost always occurs during the first note only. *Appoggiatura* is a musical term referring to the delaying of a resolution, usually by a note one step above or below the expected consonant note. The appoggiatura usually creates a dissonance, which is resolved on the following weak beat. — ED.

[39] The Picardy Grove, which was the Zinnian woods for things serendipitous, almost surely derives from the *Tierce de Picardie* ('Picardy Third'). In this musical device, the final chord of music in a minor key ends with the third of the scale being raised, thus suddenly ending the music on a major chord, rather than a minor chord. — ED.

"In a yonder world, far in the realm known to the ancients as *Extrum Yanduian Nondum Cognita*, there is much to dismay, as music has there lost its way. In the world called Earth, the natural wholes have been ripped in half. The octave of this Earth has been torn in two; its two tri-tones plead to be made whole again. Our Earth plant, we have learned, is not at peace with itself. It is sorrowful indeed!"

Other Zinnians were heard whispering similar elegies of lament. Why would there be division between people because of their chant, or the way they had life figured out, or because of their timbre? And why would Earthly presters preach hate and wreak misery upon their subjects?

"The Earth plant has lost its song!" grieved one Zinnian.

During all this reflection, the tremors and mild quakes which had precipitated the conference continued, though they did not worsen further. At the end of the session, two important questions remained :

Can Earth be rescued? And, if not, can the rest of the Universe survive without that organ?

Paramount to the first question was *why* Earth chose to hurt itself. Pliny Plankton, himself formerly a cell-person of the Earth plant, tried to explain this enigma to the crowd.

"The Earth plant itself does not perform this butchery. What you must understand is that it is individual voices — various inhabitants — of the plant which perpetrate the carnage."

But at this explication the Zinnians were growing even more baffled. How could this be, they asked. Why would an Earth person, to appease the consumptive greed of his mobile half, destroy his other half, his plant-planet, his very *orb*?

Before responding to the bewildered crowd, Pliny turned to Atollana and Ishmael, who themselves needed an explanation of the Zinnians' question.

"I'm afraid this is difficult to explain in Zinnian terms," he confided to the shrimp. "Although the Zinnians certainly know hardship and pain — indeed, they acknowledge them as part of life — they do not possess the concept of *deliberately inflicted* suffering. They know suffering only as an unwitting partner of life and beauty, from which one should free all creatures to the greatest extent possible."

Ondew nodded her head and squeezed her lips. *Yes*, she thought to herself, *when you paint one, the other gets painted with it.*

Pliny then faced his confused subjects once again and tried to elucidate the bizarre Earth condition as best he could.

"It is a disease," he told them bluntly. "All the horrors you have heard are diseases. As you know a disease will destroy its host, and in doing so destroy itself. That is what is happening to Earth." The Paradise which was Zinnia accepted sorrow as part of the dust from which Creation had been sculpted, but had no words for malice.

During all this shocking revelation, Ondew was poring over one of Pliny Plankton's documents. It was a handsome chart of the cosmos showing the location of various stars and plants (which Pliny, being a terra firmate in chant, had consistently spelled 'planet'), and all the other organs of the Universe. She scoured it looking for the one they called 'Earth', eventually locating it far in a corner of the parchment. It was carefully gilded with gold and was accompanied by an inscription which read 'Mvndvs Earth, Terra Extra Ordinem Primvs'.

Finally she knew which place this 'Earth' was. She remembered it very well, in fact she remembered it better than any other plant, for it had been her most favored, her most sublime composition. It was the plant upon which she had joyously set her dear twin, Honua. Was this really the orb that was ailing so badly? *When I painted this Earth plant*, she thought to herself, *it was as if a force far above were inspiring me. Could I really have been the sculptor of this Earth? I wonder what other power could have enkindled my hands to seed that world —* .

The very Painter of Creation was herself humbled by the Adoration of the Earth.

During all this oratory and rumination, Pliny Plankton was busily performing a complex of cosmic calculations to postulate an answer to the second question: would the self-destruction of the Earth plant result in the demise of Zinnia and even the rest of Creation?

"According to my figures," he concluded, "if the Earth plant were to annihilate itself, its neighboring plants would endure some trauma, and in particular its moon plant would suffer gravely. Zinnia and the rest of Creation might experience transitory tremors, but then normalcy would return. Earth would purge itself from the cosmos like a malignant tumor. We would all be the sadder for the death of that part of us, the Earth part, but Zinnia would not be otherwise harmed, and Creation would live on as before."

Suddenly the discussion stopped as everyone hushed to listen to the *terra*. The tremors had definitely begun to subside. They were less severe and were striking at more distant intervals. The Earth plant had apparently now moved far enough away from its precise alignment with Yandu and the Center so that its disease would not disrupt the cosmic stability of Zinnia for another year. All were relieved that any immediate danger to Zinnia had eased, but clearly the Earth crisis had to be addressed.

Prester Prawn motioned to speak :

"One could easily speculate," he proclaimed, "that in future years we may be tempted to extend Fugafest another day in order to camouflage the Earth-induced tremors. But I say that here and now we decree otherwise, that we all agree never to use Fugafest to conceal these hurtings. We should not allow ourselves to become accepting or ignorant of the sorrowful Earth pain."

There seemed to be universal enthusiasm among the Zinnians for this healthy proposal, and so it was entered into the Archives by Pliny Plankton.

There now remained the first question: was there anything that Zinnians could do to rescue their fellow Earth plant from its malady? For this inquiry, Pliny cautioned, charts and abacus could not help supply an answer. As he warned of the difficulty of this matter, an elderly Zinnian, a llama person, chanted higher to offer an opinion to the crowd :

"While we are exploring remedies for the Earth malignancy," he chanted, "perhaps we can sow a seed of the Earth plant here on Zinnia, so that — should the worst come to pass, should we lose our Earth — we would still have a young seedling of it to grow and chant the Earth song."

Here Ondew, by this point the finest Earth authority of the Intrum Yanduian bunch (that is, everyone there except those five originally from Earth), tried to clarify the admittedly naïve Zinnian comprehension of the Earth plant. Since she understood the Zinnian perspective well, she was better able to explain this than the Earth-reared folk.

"Earth . . ," Ondew then hesitated as she tried to mold her phrase. "Earth is a complex plant, a splendid, living rock garden. It is perhaps the most exquisite evolution of the painted world, but also the most vulnerable. If the creatures of Earth lose their chant, it could disrupt the entire plant. Like our isle-cosm, Earth is the sum of many creatures. The uncountable multitudes whose interwoven and interdependent existences comprise the Earth rock garden make it into a breathing plant, a plant which loves and chants like any other. But it is not actually possible to plant an Earth seedling."

She then turned to her Earth friends to be sure that she had expressed the situation accurately, about which she was not at all confident. The Earth veterans nodded in agreement, except for Honua, who smiled at her sister, put one arm around hers, faced the crowd, and interjected her own sentiment.

"Though Earth is my home and my very soul, I scarcely understand the intricacies of the Earth plant," she began. "Yet I find the idea of planting an Earth seedling here in Zinnia very

The Crustacean Codex

beautiful. And perhaps it is not altogether a futile thought. Remember that our friends the two shrimp — as well as myself and Pliny Plankton — are at home in the water seas and water rivers. And remember that we have here in Zinnia, on the far side of the Center, a sea from the Earth, the Mar Kopólo."

IX
The Sea of Kopólo

It was quite a shock to the shrimp to learn that the legendary Mar Kopólo, whose whereabouts had been lost to modern shrimpdom, was in fact right here on Zinnia. But how could this be? Pliny Plankton explained this relatively modern historical quirk to the Earth shrimp.

Apparently, the Mar Kopólo had become disenchanted (in this case meaning 'disillusioned') with the goings-on on Earth, and so all the water of that great sea changed itself into dew, evaporating all the way to Zinnia. However, unlike the dew that daily travels to Yandu, bringing with it the light and warmth of dawn, the dew from the Mar Kopólo settled intact, as water, in a lowland area of the far side of Outer Zinnia, in effect reconstituting itself there as its former sea-self.

(Whether or not this sea was actually still the same Mar Kopólo, or whether it was now a new sea even though it consisted of the very same physical water, was a great philosophical riddle which was the focus of some of the more intellectual Zinnian chants. Ishmael commented to Atollana that, on Earth, the opposite riddle occurred. On Earth, the water in the seas and oceans was always changing because it was constantly flowing into other oceans, and evaporating into clouds which then floated about and rained down into other, distant rivers and seas; yet these seas are never considered to have become *different* seas. Over the course of time a given drop of water may have been a part of every sea, however great or small, flowed through every stream, glistened in every lake, and been part of clouds blown over every continent and untold islands. So, regarding Zinnian musings over the Mar Kopólo, Atollana wondered if perhaps 'it is the same sea' and 'it is no longer the same sea' could co-exist as poetic facts, opposite but true. Ishmael commented on what a fine Yanduian tea-time topic this would make.)

What remained after this philosophical diversion was the realization that the planting of an Earth seedling in Zinnia, if admittedly the most humble of seedlings, was in fact a credible option. After much discussion, the conference adjourned with the recommendation that these four Earth-oriented people embark on an expedition to the Mar Kopólo. The shores of the transplanted sea would provide a peaceful retreat for them to decide whether they would return to Earth or settle anew on Kopólo. It would also free them from the influence of the Zinnians (who would all be pleased to have them stay in Zinnia).

Pliny noted it in the Archives as follows :

> 'Thus it was proposed this day that the Sea of Kopólo be harnessed as an indigenous substrate in which to cultivate an Earth plant culture'.

Later that day, however, reviewing his notes, he revised the clinical and ambiguous entry to read :

> 'This day it was proposed that the Mar Kopólo, an Earth sea garden living safely in Zinnia, be considered to nurture the Earth seed anew, beginning with four Earth sproutlings, being the two Earth shrimp, Honua, and Pliny Plankton, the present chronicler'.

The Zinnians themselves, it was agreed, could do little to assist the Earth plant, except to keep their own twig of the Universe, Zinnia, as healthy and full of beautiful chant as possible.

The Sea of Kopólo

Ondew and Prester Prawn were the odd people in this equation. Although Prester Prawn was originally an Earth person, he had been the benevolent despotic figurehead of Zinnia for so long that his roots had (figuratively, of course) grown into Zinnian soil. He would do his best to keep Zinnia apprised of the plight of their brethren on the Earth plant, as well as to notify them if in any way the positive aspects of Zinnian existence might somehow be chanted over to Earth. And Ondew, ironically, knew little of the painted world and its troubles. She would return to her garden jungle on the 'top' side of the Middle, where her presence nurtured the soul of all Creation. Her influence was so abstract that it was torturously frustrating for her, but that was where she was needed, and that was where she knew she must stay. The urgency of the Earth trauma having made her want to return soon, she hugged the two shrimp, Pliny Plankton, and of course her sister, Honua.

"I pray," she told them (had the gathering at large known that Ondew was the very Sculptor of Creation, they would have wondered to whom she could pray), "that someday I will no longer feel the Earth tremors, and that you will all come swimming through the ocean sea to visit me."

Ondew bowed her head respectfully to all of Zinnia and walked to the bo trees. She was returning home through the Center where, perhaps, she might first pass some time peacefully reflecting on the events of the past few days before emerging on the other side in her own island-world.

After Ondew slipped past the bo trunks and into Nothingness, the Earth people immediately began preparations for their crossing to Kopólo. The Mar Kopólo had settled in a basin on the far side of the Center, that is, past the Aqua Abyss on the opposite end from Yandu and from which Ondew and the shrimp had reached Zinnia. Atollana unrolled one of Pliny Plankton's charts and indeed located it there, being designated '*Mer des Kopólo*'. It appeared to be about a three-day journey from their present location in Inner Zinnia.

Many who lived in the sparsely populated far side of Zinnia, especially those with young children, had returned home as soon as the conference ended. But many more were still about, having stayed in case they could assist the Earth song in some way. These Zinnians now led the way for Honua, Pliny Plankton, and the shrimp. Although Pliny had once visited the transplanted Mar Kopólo, and so would be able to guide the expedition there, he tended to get lost within Zinnia proper.

The crowd gradually thinned as Zinnians reached their homes, or paths leading to them. Eventually, there was only one Zinnian left in the entourage, a very kindly deer person, who like many from the outlying regions, generally preferred to walk about on the ground. Soon he too reached his dwelling. It lay by the Zinnian steppes, at the very threshold of the Zinnian outlands.

Like most Zinnian huts, this one was thatched, but unlike those of the central regions it was well-caulked with a mud stucco to keep the warmth in, as nights in the outlands were often exceedingly cool. At the center of its roof was a round hole through which smoke rose.

"You really should come stay with us for the remainder of the night," he beseeched the four bound for Kopólo as he walked over to the entrance of the hut. "Outer Zinnia can get quite chilly at night, as you know, and all the worry over the Earth ailment has let you forget how truly tired you really are. Come inside with me. Your mission to Kopólo is a sobering matter, and you must be well-rested and mindful of the terrain. It would be better if you did not venture into the outback until morning."

Pliny Plankton, Honua, Ishmael, and Atollana all agreed that the deer spoke wisely. They thanked him and entered the friendly-looking hut. Inside, a young fawn tended the little fire whose smoke they had seen escaping through the vent in the roof.

"Grandpa!" the child cried as they entered.

"*Ragazza,*" responded the elder deer, hoping that the girl would not ask about the conference, "these four guests will stay with us this evening."

Once everyone was introduced and salutations were exchanged, the deer child quickly sought the elder deer's attention again.

"But grandfather, what happened at the meeting? Why did the ground tremble after the Fugafest was already over?" she queried.

Her grandfather was quiet for a moment, then said, "You are old enough to know the truth. Tomorrow I will sit down and explain everything as I understand it, as best I can. You see, there is a limb of the Universe called Earth. And this Earth is in great pain. Its creatures have turned against their plant and against each other. The Earth plant has — has lost its song. In the morning we will talk."

The girl's eyes tightened, probing her grandfather's face for some clue to such madness. Without speaking further she prepared a space for the guests on the floor near the fire, then quickly stole to a little alcove where she lay down to sleep. Honua, later glancing over to her, saw her gazing into nowhere, tears rolling from her eyes.

Crossing The Outlands

Pliny awoke shortly before dawn. He quietly prepared both Taprobana and Molucca tea for everyone in the hut. The deer child was the first to join him. With weary eyes she thanked him for the tea, noting with pleasure that he had spiced it with cloves, a scarce herb in this part of Zinnia.

"Are you going to the Earth plant to help it find its song?" she asked Pliny while facing the window, her eyes fixed on the first kernel of pre-dawn light budding from the outlying steppes.

"If only it were possible to find its song," he answered. "But the Earth song is a fugue more complex than any you or I could imagine. We may instead try to sing a simple Earth song in the Mar Kopólo. I have not yet decided what I will do, nor, I think, has any of the others."

Soon the nearby scrubby bushes, awakened by the modest daybreak warmth of the Zinnian outlands, began to chant a morning ode, which helped prod Honua and the shrimp from their slumber. It was loosely concocted in the style of a *passacaglia*, endowing it with a quality of great breadth and majesty.[40] Although Atollana mistook the psalm to be her dream, so that the division between the realm of her dream and that of the real world was momentarily blurred, the stimulating aroma of the cloves soon brought her fully into the awake world with the others. Now only the elder deer was not at the fire.

"Where is your grandpa?" Atollana asked the young deer.

"He often goes outside to view the outlands before dawn. He likes to visit a pass which overlooks the outlying steppes, not far from here. I think I see him coming back now."

Before the dawn burned much brighter, the elder deer entered the hut and took his place at the fire. He faced his granddaughter, who looked eager to speak.

"Grandfather, I wish to go to the Earth plant to help it find its song," the youth said.

Everyone looked at each other in surprise, and the grandfather tried to find the words to deter her. "No, no, my dear, the Earth plant needs to find its song by itself. No one else can find one's own song. Today I will explain what I learned at the conference, and what I have said will make more sense to you."

Honua, who was sitting next to the deer youth, leaned toward her and offered another thought.

---

[40] In music, the *passacaglia* form is characterized by a repeating bass line or harmonic sequence upon which the music is constructed. Closely related to the *chaconne*. — ED.

"You are already helping the Earth plant. You are a good and loving person, a beautiful soul. That is what is important. I will take your chant with me and sing it to the Earth plant. Your song will become part of the great Earth fugue."

The elder deer put his arm around his granddaughter and spoke to Pliny as the fawn studied Honua's eyes closely and nodded.

"You should begin the crossing to Kopólo soon to best pace yourselves," he suggested. "If you set out now, by nightfall you will reach an encampment which is roughly half way to the Aqua-Abyss. It lies within a silversword grove. You can usually hear the silversword chant from a fair distance. Theirs is easy to recognize. It is a very ethereal chant, like a whisper of wind. Then with daybreak you will see a high, lone mountain pinnacle in the distance. Follow that peak, for it lies roughly in line with a bridge of coral which spans the Abyss. But you need not continue any further that day. Better to stay in Inner Zinnia for the night. At daybreak the next morning, cross the bridge over the Abyss and into Outer Zinnia."

The deer glanced at the chart Pliny had brought, which Ishmael and the fawn were studying, and finished his remarks. "I see from your *carte* that you will be close enough to the route you followed on your prior trip to Kopólo to recognize major landmarks once you cross into Outer Zinnia. You should reach Kopólo by tomorrow evening."

The deer then sent his granddaughter to prepare some provisions for the four pilgrims. Once she was out of earshot the old deer explained that she had lost her parents when she was a youngster. It was a terrible hurt for her, and she has often favored Lydian chant since then (Honua reminded the shrimp of the Lydian's innate tri-tone and how the mode helps some Zinnians come to terms with sorrow).

"She endured such pain," he commented, "that she cannot bear to see any creature suffer. I knew that the news of the Earth plant would hurt her. But she is learning to accept the tragedy of her parents and all the other misfortunes which come with life. She likes to go to the Aqua-Abyss to meditate on the eternal contradiction of beauty and suffering, as it is the threshold of Outer Zinnia, the symbolic meeting of life and pain."

This was instructive for Atollana, who had crossed the Abyss by herself, with only Pliny's chart to guide her. She now understood better what the little battle scene on the chart was all about.

"The trip to the great canyon seems to help her," the elder deer continued. "Last time, she returned with sweet Ionian and Phrygian song glowing from her lips — I was so gratified to see her gentle smile again."

"We need now to reflect on all these things as well," commented Atollana, "if we are to face the Earth riddle with wisdom."

"Then you too may wish to pause at the Abyss for a day," suggested the deer. "You need not decide now, but if it would help you in planting the Earth seed in the Mar Kopólo, it would be well worth a day's delay. Perhaps the people of the Earth plant need to accept the pain that they have already suffered if they are to stop the suffering they now inflict."

The fawn returned with a canvas sack filled with food and drink for the travellers. They thanked her and her grandfather, bid them well, and set off for their crossing to Kopólo.

And so they journeyed through the outlands of the far side of Inner Zinnia, across a pass and into the Zinnian steppes, stopping only to partake of the supplies provided by the deer. Toward the end of the day, when dusk slowly enveloped the terrain and they were becoming less sure of their bearings, they heard a faint hymn, a flutey improvisation, almost like the wind whistling through a reedy grove. But there was no wind to be felt. The sound, they quickly realized, was the chant of the silversword.

They followed the woody voices of the silversword and soon came upon their modest field. The plants were singing :

> *Pilgrims, our land is rugged,*
> *Our nights quite brisk.*
> *We rarely get visitors,*
> *And our chants seldom rhyme.*
> *But warm we keep our roots*
> *And would like you to join us.*
> *Come, allow us to lure you*
> *Into a place most congenial.*

With this request, the blades of the silversword plants all began to point in one direction. The four wayfarers followed their motion to a large rock mound. On one side of this knoll was an opening which led to a concealed cave whose inner chamber was of generous size for the four travellers. They were surprised to find it quite warm inside, as within the cave a hot spring emerged from an underground channel (apparently the silversword had deliberately settled along the warmth of the subterranean stream). Honua, Pliny, Ishmael, and Atollana all jumped into the salubrious fountain to sooth their chilled and weary bodies, then sprawled out on the steamy cave floor and fell fast asleep. They had not had such a long, mellow rest since before Ondew's plunge into the Center.

In the dark of the early morning hours, Ishmael, now quite awake and unable to sleep further, ventured outside the cave. To his astonishment he found that the high rock pinnacle which the deer had mentioned could already be seen in the far distance, apparently radiating a phosphorescent glow from the Abyss. He woke the others, who upon hearing of the unexpected light all agreed to get an early start and leave for the Abyss at once. The four asked a silversword near the cave entrance (who happened to be awake) to convey their thanks to the entire grove in the morning.

So they set out, following the iridescent monolith which pierced the otherwise impenetrable black of the night outlands sky. The pinnacle appeared low above the horizon, so that light from it struck the landscape sharply at a cold, oblique angle. Thus, until dawn came, the outlands' myriad tiny features were visible to the travellers only on the surfaces which faced away from them (probably this was why the deer had not mentioned the nocturnal light as a means to travel before daybreak). They kept their pace cautious until the morning light gently puffed away the remaining charcoal veil of night.

It was only mid-afternoon when they reached the rock pinnacle. As the deer had said, this needle formation was by a coral bridge which spanned the Aqua-Abyss. The pilgrims were pleased to find that the coral had cushioned its hard, sharp self with a soft purplish moss (this was especially appreciated by the soft-skinned humans, Honua and Pliny). All four carefully walked out to the middle of the natural viaduct to afford themselves a better panorama of the magnificent canyon, which had a profound, timeless spirit, harboring little of the fog-mist which characterized it on the opposite end of Zinnia. For many, the imposing threshold symbolized the contradiction of beauty and pain.

Every soul has such a place within, a place where one's joy must somehow make peace with one's sorrow. All creatures must ultimately do this in their own way and in their own time. Even if Earth were to free itself of the hate and poison which was consuming it, the basic conflict of life and suffering would always remain. With birth comes pain, with life comes death. But whereas these were natural conflicts, the pilgrims knew that the Earth ailment was also one of malice, quite a different matter. This distinction was most important:

> Whereas life, beauty, and joy require that one ultimately make peace with pain, suffering, and death;
> Good obliges no such truce with Evil.

Such were the songs of these wayfarers. That night they camped in the open air by the rock pinnacle. The air was balmy from the warm moisture of the Abyss, and all four slept until the next day's light roused them.

The Sea of Kopólo

After finishing the last of the provisions given them by the deer, they crossed the coral bridge into Outer Zinnia and began their final trek to Kopólo. With the help of the chart he had drawn on his previous visit to the sea, Pliny recognized enough landmarks to be able to steer them through the utter barrenness of the strange peripheral landscape. Finally, he noted a range of low-lying mountains deviating only slightly from their direct course, beyond which, he said, lay the Mar Kopólo, the great lost sea which could, if they so chose, serve as a new Earth for them on Zinnia.

Upon arriving at the sea it was discovered that three elements of Earth had in fact already transplanted themselves to Zinnia. A bamboo sproutling, a swamp lily sproutling, and a lotus sproutling had evidently accompanied the Kopólean sea-dew on its exodus from Earth, for there was now a fully grown bamboo grove on one side of the Sea, many resplendent lotus plants floating on a shallow fjord near it, and an occasional white lily piercing the water. This was a surprise even to Pliny Plankton, who had seen Kopólo only once, very soon after it had first settled in Zinnia. The people of Zinnia could poetically observe that a seedling of the Earth song had already taken root in the Mar Kopólo.

Atollana, Ishmael, Pliny Plankton, and Honua all sat quietly at the edge of the Sea for a long while. They each had to choose: should they partake in the founding of a new civilization in the Mar Kopólo? Or should they return to the Earth and all its troubles, simply to live their lives as best they could, striving to be healthy cells in that troubled orb? Honua was the one eventually to break the silence.

"I left the Earth a long time ago," she began. "I left because I could not reconcile the unhappiness of those who lived without the music of their own chant, whatever it may be. On Earth they are called 'disen*chanted*' for that reason. In Zinnia I tried to find their peace. I tried and I failed, for life and music are one and the same. There is no invisible virtue to Earth's pain. Suffering is not an indelible rite of the Profane.

"Nor — I finally understood when Ondew and I faced each other alone in the perfect peace of the Center — are the Sacred and the Profane incongruous rifts of Creation. The two are in fact inextricable voices of the same fugue. On Earth this coexistence is called *mana*.[41] The Earth has *mana* because of all that have lived and loved there. If Earth is truly a life unto itself, then every nuance of every breeze, each ripple of every ocean, even the most insignificant pebble, truly shares Earth's soul."

She turned to Ishmael, who was leaning against her. "Ishmael . . . Earth is dancing with *mana*. Have we failed to nurture the great fugue which is our Earth?"

Ishmael replied that there are indeed voices nurturous of the Earth. Had not he and Atollana been lavished with love by La Vecchia, as well as by many other voices of the great Earth polyphony? Yandu, the sheep dogs' herding, the truffle hounds' snuffing, the Taprobana and Molucca plants' broth; Ondew, the fig tree's fruit, the silverswords' cave, the underground spring's warmth, and many, many other selfless kindnesses — to simply enumerate these explained naught. There was nothing in idyllic Zinnia, nothing in Ondew's islecosm, which did not also exist on Earth. What, then, was different?

The common Zinnian explanation, Ishmael continued, is that Earth has lost its song, and thus lost its link with the force of Creation. Even the old deer explained the Earth ailment to the fawn in this way. Perhaps these are not mere pretty words.

---

[41] *Mana* is the concept of an inanimate object possessing something beyond its obvious material self, i.e., possessing power, soul, energy, or magic. Also associated with animism. Mana is a common concept in non-Western cultures; the term is of Polynesian origin. The philosophy of mana is alive and well in the West but conventionally denied; e.g., an object we believe to be antique will possess special power over us that an indistinguishable reproduction will not; an otherwise mundane place where we first fell in love will assume mana for us the rest of our lives; an object (e.g., a violin, a house) owned or used by a famous person (legendary violinist, movie star) will possess a special mana beyond itself, though the object is unchanged; a work of art which moved us forcefully for years will suddenly lose its magical power when its long-standing attribution to a revered maker is discredited. — ED.

But how does one remedy such an ailment? And how tempting, instead, to nurture a young sea, to cultivate a new world with care and truth.

Atollana's eyes, which had been fixed on nearby lotus petals, now gazed upon the far shores of Kopólo. "The night La Vecchia died, we were determined to examine Aquanesia from a neutral distance," she recalled. "Though Yandu himself had left Earth in pursuit of the neutral observance of all things, he has not solved the Earth riddle. Would the Mar Kopólo allow we of Earth a dispassionate look at forgotten truths and accepted falsehoods?"

Pliny Plankton, book and quill in hand, had been immortalizing the events as always. Now, however, he put his equipment down and spoke without recording his words, as though Historie had, for the moment, been suspended.

X
Epilogue

"I am a scientist and an historian," Pliny is said to have begun. "I record history so that the future can never be ignorant of the errors of the past, nor uninspired by its virtues. I didn't want the passing of time to vindicate or glorify the world's despots, nor to forget the modest and good. Similarly, I pursue the disciplines of the sciences because science harmonizes the offerings given us by the Earth and the rest of Creation. Good science does that. Science is poetic, and like all energies, science is a form of music. Science is my chant. But on Earth some science is governed by greed rather than poetry. If I can continue the work I abandoned on Earth, if I can help expose false presters and vulgar science, perhaps I will better that world by the one tiny component cell-voice of it called Pliny Plankton. But yet, to free oneself of that inherited burden, to start anew, I could make myself a new Book and begin on Page One. What greater gift could we imagine? Here is a virgin ocean patiently awaiting outcasts or expatriates such as ourselves. Why should we not allow ourselves to be seduced by it? If we were to return home and the Earth were then to perish, we would perish with it, and the Earth seed would have lost its opportunity to impregnate itself in the Mar Kopólo."

Those were the last words heard at the *Mer des Kopólo* that fateful evening. Honua, Atollana, Ishmael, and Pliny Plankton sat silently for a long time without venturing from the bamboo grove. Dusk came, and still they sat, cuddling together to keep warm, no one commenting on any of the others' words, no one

Epilogue

saying anything further. All four had made their individual decisions.

The Zinnian/Kopolian night gave way to the first whispers of the day's rebirth. A crimson glow of early morning dew-light exuded its resplendent warmth through the lily and beamed dancing shadows off the bamboo. Two shrimp and two human people peacefully allowed the music of the Kopolian dawn to penetrate them. Before daybreak came, they fixed eyes on each other to acknowledge that the time had come for them to step foot into their chosen destinies; the wayfarers arose from the hospitable grove and swam out into the luscious, open waters of the Mar Kopólo.

All four had, of course, been in water since leaving Earth. Atollana and Ishmael had traversed the ocean sea moat, Honua often wandered Zinnian rivers, and Pliny bathed every morning in a forest lake near his hut. These were splendid, loving seas indeed. But Kopólo was a sea born of their own *terra*, virtually barren and anxious to be the womb of an Earth garden in Zinnia. Perhaps some algae, some coral, some fellow fishfolk and terra firmates might all like to join in creating a new world in this virgin ocean and on its shores. It was an inspiring idea. The Aquanesian Sea and all the other waters and terra firmas of Earth could perish, but the new settlement, evolving in an ocean and *terra* undefiled by the deprivations afflicting Earth, would be unaffected.

Conversely, each could return to their Earth, the shrimp to the Aquanesian Sea, and Pliny and Honua to whatever corner they wished, healthy cells radiating their own chant. Perhaps after a future Fugafest there would be no extra tremors in Zinnia, and the Zinnians, knowing why, would smile and chant a very happy song. Perhaps the sorrowful Earth songs that the Zinnians now chanted would eventually find peace in the serendipitous Picardy Grove. Perhaps if the citizenry and denizenry of Aquanesia could know the Universe as nothing more than chamber music, a master ensemble in which even

The Crustacean Codex

Chaos listened to all the other players, perhaps peace and beauty would return.

The compassionate glow of Zinnian daybreak danced off dew which had settled upon the Mar Kopólo during the night. The four pilgrims from Earth fixed eyes on each other, as if to acknowledge an unspoken pact. Together they swam to the middle of the Sea, their heads just piercing its surface.

Let us all hold tightly
onto the dew, whispered Honua
passionately, *let us ascend with it. Dear*
beautiful Sea of Kopólo, joyously nurture
your bamboo, your lily, your lotus.
But we shall return home,
the dew, Earth,
and we as
one.

CHIENGMAIUS, IANARIUS 2
AN•DO• MCMXCIII
abcdefghijklmnopqrstuvwxyz
1234567890

This book was written during otherwise free time
on planes, trains, buses, and in hotels, airports, and backstage
during concert tours in the Far East.
The draft was completed visiting family on a rice paddy
near Chiengmai, Thailand, in January of 1993.
— T.S.